Commercial Production of Horticultural Crops

Commercial Production or Horticultural Crops

Kunal Mitra

Oxford Book Company
Jaipur ▪ New Delhi

Published By

Oxford Book Company

Regd. Off. "Bony Residency", Gate No. 2
Opp. Tilak Public School, Vishweswariya Nagar,
Gopalpura Road, Jaipur - 302018, India
Ph: 0141 - 2761280, Telefax: 2761381
email - oxfordbookjpr@gmail.com
website - oxfordbookcompany.com

Delhi Office: 102, 1st Floor, Satyam House,
4327/3, Ansari Road, Darya Ganj,
New Delhi - 110002, India
Ph: 011-45652440

Distributed By
KEYA INTERNATIONAL PRESS LLP
E-mail: keyainternationalpressllp@gmail.com

Published in India, 2024

ISBN : 9788189473464

Price: ₹ 1975.00

Typeset by Shivangi Computers, Jaipur

Printed and bound in India by Thomson Press Ltd.,
New Delhi

Preface

Horticulture is the area of agriculture that deals with garden plant cultivation: the science and art of growing fruits, vegetables, ornamental trees, bushes and flowers. The primary aim of horticultural science is to develop plants of the highest quality that offer the promise of high yields. To achieve this goal, and to develop methods for protecting horticultural crops from the ravages of insects and diseases, many different science are involved, which make this arena a multi-disciplinary area of study, practice and research. Today, as commercialisation defines any arena of endeavour the importance of growing crops yielding high commercial value in horticulture cannot be sidelined.

Horticulture has three major commercial branches: the nursery industry that supplies young trees and shrubs for fruit growers, gardeners and landscape designers; the plant growing industry that raises greenhouse and field grown flower and vegetable plants; and the seed-producing industry that specialises in the growing of plants for the their seed. This text not only delineates for the reader the processes, principles and practices potential and scope they have, particularly with regard to economic revenue and horticultural development. In addition, it also takes into account current trends and developments in the field, incorporating critical perspectives which also touch upon the dilemmas and issues involved in raising commercial crops in horticulture.

Editor

Contents

1

Horticultural Production in India

The creditability of horticulture has been well established in improving productivity of land, generating employment's, improving economic conditions of the farmers and entrepreneurs, enhancing exports and above all providing nutritional security to the people. The sector, which includes fruits vegetables, root and tuber crops, mushroom, floriculture, medical and aromatic plants, cashewnut, plantation crops including coconut and oil palm has established its creditability. Focussed attention to horticulture was given only during Eighth Plan although tune was set in VII Plan after achieving self-sufficiency in food grains. Contribution of Horticulture to GDP of agriculture is estimated to be more than 24.5 from less than 8 percent of cultivated area.

The investment in horticulture sector with focussed attention has resulted in spectacular change in the horticultural scenario, and production and productivity of horticultural crops, which have increased many folds. India has emerged as second largest producer of fruits and vegetables, and accounts for a major share in the export of cashewnut. Exports of fruits and flowers have also increased many folds. The issue that is most clearly emerging is that the horticulture sector in the country despite numerous challenges and short comings, is moving dynamic and is in a crucial phase of development. The scenario, which has proved the potential of horticulture in agri-business, have encouraged the private sector activities and in this context, quality management has become important.

GROWTH OF HORTICULTURE

The focused attention to Horticulture has paid dividend and resulted increased production and export. Large area was brought under improved cultivars, production of quality planting material as seed increased, large number of farmers were trained, and innovative technology like drip irrigation, green house cultivation were encouraged. Consequently availability of fruits, vegetables and flowers increased. Evidently, more than 50 per cent increase in production of fruits and vegetables is seen between 1991-92 and 1999-2000. Impact of increased production of horticultural produce is also seen on volume and value of export of horticultural produce. Maximum increase in export during the period was for spices, cashew and for floriculture produce. However, increase in export of fresh fruits and vegetables was not appreciable although processed fruits and vegetables have significant increase.

FRUITS

India is the second largest producer of Fruits after China, with a production of 44.04 million tonnes of fruits from an area of 3.72 million hectares. A large variety of fruits are grown in India, of which mango, banana, citrus, guava, grape, pineapple and apple are the major ones. Apart from these, fruits like papaya, sapota, annona, phalsa, jackfruit, ber, pomegranate in tropical and sub-tropical group and peach, pear, almond, walnut, apricot and strawberry in the temperate group are also grown in a sizeable area. Although fruit is grown throughout of the country, the major fruit growing states are Maharashtra, Tamil Nadu, Karnataka, Andhra Pradesh, Bihar, Uttar Pradesh and Gujarat.

Table 1. Production of Fruits in Indai during 1999-2000

Crop	Area (000 ha.)	Production (000 MT)
Apple	238.3	1047.4
Banana	490.7	16813.5
Citrus	526.9	4650.6
Grapes	44.3	1137.8
Guava	150.9	1710.5
Litchi	56.4	433.2
Mango	1486.9	10503.5
Papaya	60.5	1666.2
Pineapple	75.5	1025.4
Sapota	64.4	800.3
Others	601.2	5707.6
Total	3796.8	45496.0

Mango is the most important fruit covering about 35 per cent of area and accounting of 22 per cent total production of total fruits in the country, which is highest in the world with India's share of about 54%. India has the richest collection of mango cultivars. Major mango growing States are Uttar Pradesh, Bihar, Andhra Pradesh, Orissa, West Bengal, Maharashtra, Gujarat, Karnataka, Kerala and Tamil Nadu. The main varieties of mango grown in the country are Alphanso, Dashehari, Langra, Fajli, Chausa, Totapuri, Neelum etc.

Banana comes next in rank occupying about 13 per cent of the total area and accounting for about 34.2 per cent of the total production of fruits. India has first position in the world in banana production. While Tamil Nadu leads other States with a share of 19.00 per cent, Maharashtra has highest productivity of 58.60 metric tonnes against India's average of 32.50 metric tonnes per ha. The other major banana growing states are Karnataka, Gujarat, Andhra Pradesh and Assam The main varieties of banana are Dwarf Cavendish, Bhusaval Keli, Basrai, Poovan, Harichhal, Nendran, Safed velchi etc. Citrus fruits rank 3rd in area and production accounting for About 12 and 10.4 per cent of the total area and production respectively.

Lime, lemons, sweet oranges and mandarin cover bulk of the area under these fruits and are grown mainly in Maharashtra, Andhra Pradesh, Karnataka, North-Eastern States, Punjab, Orissa and Madhya Pradesh.

Guava is the fourth most widely grown fruit crop in India. The area under guava is about 0.15 Million ha producing 1.80 MT. The popular varieties of guava are Allahabad Safeda, Lucknoe-49, Nagpur Seedless, Dharwar etc. Bihar is the leading state in guava production with 0.30 MT followed by Andhra Pradesh, and Uttar Pradesh.The other states where guava is grown widely Gujarat, Karnataka, Punjab and Tamil Nadu.

Grapes occupies fifth position amongst fruit crops with a production of 1.08 MT from an area of 0.04 Million ha. The major varieties of grapes grown in India are, Thomson Seedless, Sonaka, Anab-e-Shahi, Perlette, Banglore blue, Pusa seedless, Beauty seedless etc. Maharashtra occupies the first position with a production of 0.68 MT of grapes, followed by Karnataka. The other states growing grapes are Punjab, Andhra Pradesh and Tamil Nadu. The other major fruits grown in the country are Apple, Litchi, Papaya, Pineapple and Sapota

VEGETABLES

In vegetables production, India is next only to China with an annual production of 87.53 million tonnes from 5.86 million hectares having a share of 14.4 per cent to the world production. Adoption of high yielding cultivars and FI hybrids and suitable production technologies has largely contributed for higher production and productivity. Per capital consumption has also increased from 95 gram to 175 gram per day. More than 40 kinds of vegetables belonging to different groups, namely cucurbits, cole crops, solanaceous, root and leafy vegetables, are grown in different agro-climatic situations of the country. Except a few, namely brinjal (egg plant), colocasia, cucumber, ridge gourd, sponge gourd, pointed gourd etc., most of the other vegetables have been introduced from abroad.

Potato is most widely grown vegetable crop in the country with a share of 25.7 per cent. The area under potato cultivation is 1.28 Million ha with total production of 22.49 MT. The main varieties of potato grown in the country are Kufri Chandramukhi, Kufri Jyoti, Kufri Badshah, Kufri Himalani, Kufri Sindhuri, Kufri Lalima etc. Uttar Pradesh is the leading potato growing state in the country with a production of 9.53 million tonnes followed by West Bengal and Bihar.

Tomato occupies second position amongst the vegetable crops in terms of production. The total production of tomato in the country in 1998-99 was 8.27 MT from an area of 0.46 M. ha. The main varieties of tomato grown in the country are Pusa Ruby, Pusa Early Dwarf, Arka Abha, Arka Alok, Pant Bahar, Pusa hybrid-1, Pusa hybrid-2, MTH-6, Arka Vardan etc. Andhra Pradesh is the largest grower of tomato with a production of 2.05 MT. The other main tomato growing states are Bihar, Karnataka, Maharashtra and Orissa.

Brinjal occupies the third position amongst vegetable crops. The production of brinjal in the year 1998-99 was 7.88 MT from an area 0.49 M.ha. The varieties of brinjal popular in the country are Arka Navneet, Pusa Ankur, Hybrid-6, Pusa hybrid-5, ARBH-1, ABH-1, Pusa Purple Long, Pusa Purple Cluster, Ritu Raj etc. West Bengal is the largest producer of brinjal followed by Maharashtra and Bihar. The other main state growing brinjal Karnataka, Maharashtra, Gujarat, Andhra Pradesh, Assam and Madhya Pradesh. Cabbage is the fourth most widely grown vegetable crop of our country. India is the leading country producing Cabbage. The area under Cabbage cultivation is 0.23 M.ha producing 5.62 MT. The main varieties of cabbage are Pusa

Drum Head, Golden Acre, Pride of India, Pusa Mukta, Pusa Synthetic etc. West Bengal produces 1.84 MT and is the largest grower of the cabbage. Orissa and Bihar occupies second and third position respectively. The other major growers of cabbage are Assam, Karnataka, Maharashtra and Gujarat. The other important vegetable crops grown in the country are onion, chillies, peas, beans, okra, cabbage, cauliflower, pumpkin, bottlegourd, cucumber, watermelon, palak, methi, carrot and radish.

FLORICULTURE

Commercial floriculture has been of recent origin though the traditional flower cultivation has been going on for centuries. Emphasis has been shifting from traditional flowers to cut flowers for export purposes. The liberalisation of economy since 1991-92 has given an impetus to the Indian entrepreneurs for establishing export oriental floriculture unit under controlled climatic conditions. Maharashtra, Karnataka, Andhra Pradesh and Haryana have emerged as major floriculture centres in recent times. At present the total area under floriculture cultivation is about 73,970 ha. with a production of 4.60 lakh tonnes of loose flowers and 1155 lakh (numbers) of cut flowers. Exports of floriculture products valued at Rs. 100 crores. There has been an impressive growth in the export of cut flowers from Rs. 1 crores in 1987-88 to 100 crores in 1998. More than 170 EOUs have been approved in the sector out of which 70 units are operational. But many of them operate at less than 50 per cent of their capacity. With the growing competitiveness, floriculture units in India have been facing several constraints and with redressal of the problems, the industry may turn to be viable enterprise for earning foreign exchange.

SPICES

India, known as "Land of Spices", is the largest producer, consumer and exporter of variety of spices in the world. The area covered under various spices in the country is estimated to be 25.17 lakhs ha with an annual production of 29.10 lakhs tonnes. More than 90% of the spices produced in the country is used for domestic consumption and the rest exported as raw as well as value added products. The important spices produced in the country are: Black pepper, ginger, turmeric, garlic, chillies, coriander, cumin, fennel, fenugreek, celery, clove, cassia, nutmeg, mace, cardamom, saffron, vanilla and a group of herbal spices.

Chillies occupies the top position amongst spices with a share of 30 per cent. Total production of Chillies in the year 1998-99 was The share of spices in the total agricultural export during 1998-99 was about 6% with an export of 2.31 lakh tonnes earning foreign exchange worth Rs. 1758 crores. The exports of spices and spice products during 1999-2000 was 2.09 lakh tonnes valuing Rs. 1861 crores. Pepper was the leader in export earning with 46% share followed by Oil & Oleoresins (15%), chillies (13%) and turmeric (6%).

MEDICINAL AND AROMATIC PLANTS

India has been considered as treasure house of valuable medicinal and aromatic plant species. Ministry of Environment and Forests have identified and documented over 9500 plant species considering their importance in the pharmaceutical industry. In the present context of 'back to nature' in health care, it is relevant that these valuable plant species are not only preserved but also their cultivation developed in order to meet the entire demand of the domestic industries as also to exploit the bright prospect for export. Shift from collection to cultivation of medicinal & aromatic plants will also ensure purity, authenticity and sustainable supply of raw materials required for herbal drugs, including polyherbals. Our foreign exchange earning potential from these group of plants is estimated to be over 3000 million US dollars per annum. Agro-techniques have been developed for large number of medicinal plants by the State Agricultural Universities. Due to unorganised marketing arrangements this sector has not exploited the full potential. A Medicinal Plants Board has been constituted in the Department of Indian Systems of Medicines & Homeopathy to address all the issues.

PLANTATION CROPS

Coconut

The coconut palm eulogized as " Kalpavriksha " is one of the most valuable gifts of nature to mankind. It serves as a veritable source of food and drink to million of people in the tropics. The crop is grown mainly all along the coastal parts of the country and to some extent in the interior parts and also in the North Eastern regions. The important coconut growing states in India are Kerala, Tamil Nadu, Karnataka, Andhra Pradesh, Orissa, Maharashtra, Goa, Assam, Pondicherry, Lakshadweep, A&N Islands. Other states like Gujarat, Madhya Pradesh, Bihar, and North Eastern regions have also gained momentum in coconut cultivation. The crop assumes considerable significance in national economy in view of its vast opportunities for employment and income generation. Major share of coconut production is contributed by millions of small and marginal farmers. About 10 million people are depending on coconut cultivation, processing, and related activities

Cashewnut

In the world scenario, India occupies a premier position contributing about 43% of the cashewnut production. Despite India being the largest producer and exporter of cashewnut, the production of raw cashewnut in the country is far below the requirement of the processing sector. The maximum production of raw cashewnut was 5.2 lakh tonnes obtained from an area of 6.86 lakh ha. during 1999-2000. This production is hardly sufficient to meet 50% of the demand of about 825 cashew processing units in the country. The average demand growth rate for kernels for export and internal consumption is around 13% per annum. The demand of the industry is about one million tonnes of raw cashewnut.

Cocoa

About 14000 ha are under cocoa cultivation in India. Due to sudden crash in the prices in the eighties, there has been a decline in the area. In 1999-2000, 5200 tonnes of cocoa was produced. Kerala accounts for the bulk of the production. Cocoa is mainly used as a confectionery item. The annual export earnings from cocoa is to the tune Rs. 9 crore. Compared to the global production of 29 lakh MT, the Indian contribution is insignificant. To fill the gap between the production and requirement intensive programme for production need to be initiated.

ORGANISATIONAL SUPPORT

The Department of Agriculture and Cooperation in the Ministry of Agriculture is headed by Secretary (Agriculture & Cooperation) assisted by horticulture Commissioner is vested with the responsibility for the development of horticulture and it is a nodal agency for horticulture development. The Horticulture Commissioner heads the Division of Horticulture and the division has two autonomous organisation and two attached office and also attached a National Committee on Use of Plastics in Agriculture. Besides, programme are being implemented by State Department of Agriculture, horticulture and state Agricultural Universities and also Institutes of ICAR. In addition, linkages have also been established with organisations of other ministries for the development of horticulture i.e. APEDA, Coir Board, Spices Board etc.

The Horticulture Division in the Department of Agriculture and Co-operation, vested with the responsibility of overseeing the overall accelerated development in the national perspective has the mandate to:

- Implement the programmes for improving the production, productivity and utilisation of horticultural crops.
- Support and formulate policies aimed for accelerated growth of horticulture.
- Facilitate the availability of disease free planting material and seeds of horticultural crops.
- Provide the leadership and co-ordinate activities for the promotion of horticulture.
- Work as facilitator for the transfer of technology to farmers and promote the use of information technology.
- Promote better utilisation and increased consumption of horticultural produce to ensure higher returns to farmers and nutritional security to people.
- Develop strong base for the supply of inputs, transfer of technology and human resource developmental activities.
- Promote horticulture in NE Region, hills, tribal & backward areas for improving economic status of people.

The Division has the support of two Boards, two Directorates and a Committee. A brief description about their activities are given below:

Coconut Development Board (CDB)

The Coconut Development Board, established in 1981 has a mandate for integrated development of coconut industry in the country through promoting production, processing, marketing and product diversification of coconut. The functions of the Board are: adopting measures for the development of coconut industry, recommending measures for improving marketing of coconut and its products, imparting technical advice to those engaged in coconut cultivation and industry, provide financial and other assistance for expansion of area improving productivity and product diversification, compile statistics on coconut, undertake publicity and publication activities etc. Encouraging adoption of modern technologies for processing of coconut and its product, recommending measures for regulating imports and exports for coconut and its products, fixing grades, specifications and standards for coconut and its products also falls under its ambit.

The Board has its Headquarters at Kochi, Kerala, with three regional offices at Bangalore, Patna and Chennai and six state centres at Hyderabad, Calcutta, Guwahati, Agartala, Bhubaneswar and Port Blair. In addition, Nine Demonstration cum Seed Production Farms have also been established in the States of Kerala, Karnataka, Andhra Pradesh, Orissa, Assam, Bihar, Tripura and Madhya Pradesh. The Board has been implementing various development programmes for the last 19 years.

National Horticulture Board (NHB)

National Horticulture Board was set up by the Government of India in 1984 as an autonomous society under the Societies Registration Act 1860 with its Headquarters at Gurgaon, Haryana with the mandate to promote integrated development of horticulture, to help in coordinating, stimulating and sustaining the production and post-harvest management of fruits and vegetables and other horticultural crops. NHB has the following mandates:

a) To develop high quality horticultural farms in identified belts and make such areas vibrant with horticultural activity which in turn will act as hub for developing commercial horticulture
b) To develop post-harvest management infrastructure

c) To Strengthen Market Information System and horticulture database

d) To Assist R&D programmes to develop products suited for specific varieties with improved methods and horticulture technology

e) To Provide training and education to farmers and processing industry personnel for improving agronomic practices and new technologies and

f) To Promote consumption of fruits/vegetables in fresh, processed form etc.

NHB has network of 33 regional offices located through the entire length and breadth of the country to promote horticulture in the country. The initiatives of the Board has helped in developing infrastructure for horticultural development.

Directorate of Arecanut & Spices Development (DASD)

The Government of India established the Directorate of Cocoa, Arecanut & Spices Development at Calicut, Kerala with effect from 01.04.1966 as one of the subordinate office of the Ministry of Agriculture (Department of Agriculture & Cooperation). The Directorate has the mandate to formulate appropriate development schemes on spices, medicinal & aromatic plants and arecanut at the national level. These schemes are implemented through State Governments, Agricultural Universities, ICAR Institutes as well as through Regional Research Laboratories under CSIR and monitored by the Directorate. This Directorate also collect, compile and publish data on area, production, price trends, export and import of arecanut and spices in addition to keeping in liaison with research and development agencies at the State and Central levels. Through the efforts of Directorate there has been unprecedented increase in production and quality of these products.

Directorate of Cashewnut & Cocoa Development (DCCD)

The Directorate of Cashewnut Development came into being with effect from 01.04.1966 at Cochin as subordinate office of the Union Ministry of Agriculture with primary objective of development of cashewnut in the country. Development of cocoa was transferred to this Directorate during 1998. The Directorate of Cashewnut and Cocoa Development is responsible for formulation and coordination of schemes programmes for development of cashewnut & cocoa in the country. The Directorate maintains close liaison with the State Governments and other State level agencies in the course of development.

The Directorate conducts comprehensive studies on various aspects of cashewnut & cocoa development including production, prices, marketing and other related problems. It has been instrumental in dissemination of technical information and research findings of practical value through publications. Institute of Directorate has helped in adoption of improved technology for production and availability of assured quality planting material. The Directorate also publishes its journal regularly.

National Committee on Plasticulture Applications in Horticulture (NCPAH)

The Committee is responsible for horticulture development through plasticulture intervention, which was initially started in the Ministry of Chemical and Fertilizer and is attached to this ministry since 1993. NCPAH has 17 Precision Farming Development Centres (PFDC) working for popularization of precision farming and hi-tech horticulture applications like microirrigation, green house cultivation, mulching etc. to achieve improved production. Union Minister of Agriculture is the Chairman of NCPAH and the Horticulture Commissioner is the Member Secretary.

Beekeeping Development Board

This is a coordinating Board to integrate the programme on honey bee. The Board is Headed by Secretary (Agriculture & Cooperation). Horticulture Commissioner is Member Secretary to the Board. The Board decides upon various programme to be taken up for the development of honey bee. Besides the above, support for the development especially for marketing and export is also provided through Agriculture Produce Export Development Authority and Spice Board in the Ministry of Commerce.

Schemes for Human Resource Development in Horticulture

The objective of the scheme is to promote rapid growth of horticulture by bridging the gap of knowledge and skill both managerial and technical by training people to become entrepreneurs, or self-employed in the horticulture sector and to create skills for employability in the horticulture units/farms and upgrade the knowledge of departmental staff in the field of horticulture.

Pattern of Assistance:

— Promoting the rapid growth of horticulture by bridging the gap of knowledge and skill both managerial and technical by training people to become entrepreneurs or self-employed in the horticulture sector.

— Create skills for employment in the horticulture units / farms and upgrade the knowledge of departmental staff in the field.

— Impart practical training at different levels i.e. training of supervisors; training of gardeners; training of entrepreneurs and training of departmental staff.

— Infusion of technology competent human resource for the development of horticulture

— Technical upgradation of the people concerned with horticultural development

The scheme will have four components:

a) Training of Supervisors
b) Training of Gardeners
c) Training of entrepreneurs
d) Training of departmental staff.

The participants to these courses will be invited through open advertisements in local newspapers. The minimum qualification for Supervisory & entrepreneurs training programme is Higher Secondary and for Gardeners, Class-VIII(Middle)standard. States, which already have a strong institutional set up for training in horticulture, will only be included in the scheme, in the initial stages. The institutions identified for training will already have the basic infrastructural facilities like class rooms, staff, hostel facilities etc. Having regard to this fact, SAUs, KVKs, Mali training centres, Zonal training centres, Institutions run by NGO's etc will be selected for giving training under the scheme. The maximum number of institutions in each states will not exceed two to three for both training components. Based on the willingness of the participating Institutions, proposals will be invited directly from these institutions and funds released to them.

Under the schemes, the institutions will be provided funds for the following:

a) *Infrastructure*: Addition to class rooms and hostel facilities; training , setting up nurseries/gardens for demonstration purposes; gardening equipment's and tools etc.

b) *Operational costs*: incremental costs towards payments to temporary staff; payments to guest lecturers; payment of stipends to the participants; TA/DA incidental to the training, hiring of vehicle for field visit of trainees, contingency etc.

The courses are intended to be of one year duration for Supervisors and of six months for gardeners and of three months for Entrepreneurs. In order to attract the candidates and more importantly retain them and prevent their drop out, a monthly stipend will be provided to them in the form of Boarding & Lodging in the concerned Institute. The courses will be residential. At the end of the training Supervisors will be awarded a Diploma in Horticulture and Gardeners will be awarded a certificate. These trained people would be the potential candidates for employment by the Firms engaged in horticulture development. Similarly, about 15 to 20 in-service personnel will be trained in the local institutes. For training of trainers and in-service personnel, each state will maintain a panel of candidates. The candidates would be nominated for training based on the training programme which are announced by the Training Institutes. Candidates, particularly Trainers who are required to train others could be deputed for training abroad also, to limited extent, for which assistance would be made available for meeting the travel cost and course fee. The concerned State Departments of Horticulture/ Agriculture would function as the nodal department for this purpose. Funds would be made available to the State Horticulture/Agriculture Departments for meeting the expenses of the Departmental candidates based on their specific proposal. In case the States wish to organise specialized training courses on horticulture related subjects, assistance will be provided for the same to the concerned Institutes directly based on the recommendation of the concerned states. Such training would generally be of short duration of 7-10 days for 20 to 25 participants.

REFERENCES

Chadha, K.L., Ravindran P.N. and Leela Sahijram, *Biotechnology in Horticultural and Plantation Crops*, New Delhi, Malhotra Publishing, 2000.

Janick, Jules. 1979. *Horticultural science*. San Francisco: W.H. Freeman. Page 1.

Saroj, P.L. et.al., *Advances in Arid Horticulture: Present Status*, International Book Distributing Co., Lucknow, 2004.

2

Production of Tropical and Sub-tropical Fruits

BANANA

Banana (*Musa paradisiaca L.*) occupies over 1,64,000 hectares, mainly in Tamil Nadu, West Bengal, Kerala, Maharashtra, Gujarat, Karnataka, Assam, Andhra Pradesh and Bihar. Though some inferior types of banana are found growing as far north as the Himalayas, its commercial importance is mainly limited to the more tropical conditions, such as those prevailing in central, southern and north-eastern India. It is a moisture- and heat-loving plant and cannot tolerate frost or arid conditions.

Varieties

Cultivated varieties are broadly divided into two groups: table and culinary. Among the former are 'Poovan' in Madras (also known as 'Karpura Chakkarekeli' in Andhra Pradesh); 'Mortaman', 'Champa' and 'Amrit Sagar' in West Bengal; 'Basrai', Safed Velchi', Lal Velchi' and 'Rajeli' in Maharashtra; 'Champa' and 'Mortaman' in Assam and Orissa; and 'Rastali', 'Sirumalai', 'Chakkarekeli', 'Ney Poovan', 'Kadali' and 'Pacha Nadan' in southern India. 'Basrai', which is known under different names, viz. 'Mauritius', 'Vamankeli', 'Cavendish', 'Governor', 'Harichal', is also grown in central and southern India.

Recently, the 'Robusta' variety is gaining popularity in Tamil Nadu and Karnataka. The 'Virupakshi' variety (Hill banana) is the most predominant variety in the Palni Hills of Tamil Nadu. Among the culinary varieties, Nendran bananas, 'Monthan', 'Myndoli' and 'Pacha Montha Bathis' are the leading commercial varieties in southern India. 'Gros Michel' is a recent introduction into southern India; it is suitable for cultivation only under garden-land conditions and is generally fastidious in its cultural requirements. It is not, therefore, in favour with the cultivation.

Propagation and Planting

Propogation is by suckers or off-shoots which spring at the base of a banana-tree from underground rhizomes. Vigorous suckers, with stout base, tapering towards the top and possessing narrow leaves, are selected for plant. Each sucker should have a piece of underground stem with a few roots attached to it.

Banana suckers can be planted throughout the year in southern India, except during summer, whereas in the rest of the country, the rainy season is preferred. They are planted in small pits, each just enough to accommodate the base of a sucker. The planting-distance varies from 2m X 2m in the case of dwarf varieties to 4m X 4m in the case of very tall varieties.

Manuring

An application of 20 to 25 kg of farmyard manure, together with about 5 kg of wood-ashes per plant is given at planting time. In southern India, ammonium sulphate is applied one month, five months and nine months after planting 20 kg per hectare each time. In western India, a little over 2 kg of oilcake per stool is applied during the first three months after planting. A complete fertilizer mixture may be applied to supply 100 to 200 kg of N, 100 to 200 kg of P_2O_5 and 200 to 400 kg of K_2O per hectare. A green-manure crop is also considered beneficial. Trials at the Indian Institute of Horticultural Research have shown that for the 'Robusta' variety, a fertilizer mixture comprising 180 g of N + 108 g of P_2O_5 + 225 g of K_2O per plant is ideal.

After-care

The removal of suckers, dry leaves and pseudostems, from which the fruits have been harvested, constitute the main after-care. Daughter-suckers should be removed promptly until the mother-plant flowers, when one daughter-sucker may be allowed to take its place. The removal of dry leaves and useless pseudostems requires to be done in time. After all the fruits are formed, the pendant portion of the remaining inflorescence along with the heart should be removed.

The propping of plants with bamboo poles, especially those which have thrown out bunches, is necessary wherever damage by wind is apprehended. Where the wind damage is recurring, dwarf varieties should be preferred.

Irrigation

The banana-plants require very heavy irrigation. Irrigation is given in most places once in seven to ten days. Stagnation of water in the soils is not very congenial to the proper growth of banana and, hence, the drainage of soil is also essential.

Harvesting

Early varieties commence flowering in southern and western India about seven months after planting, and the fruits take about three months more to ripen. In the Andhra Pradesh delta areas, the fruits are ready for harvesting about seven to eight months after planting. The first crop of the 'Poovan' variety matures in 12 to 14 months and the second in 21 to 24 months after planting. In other parts of India, the first crop is usually gathered a year after planting, whereas the succeeding crop may be ready in six to ten months thereafter.

The bunch is harvested just before it attains the ripening stage. When the fruits have reached the full size, they become plump, and mature with a distinct change in colour. For long transport, the bunch may be harvested somewhat earlier. The bunch is cut, retaining about 15 cm of the stem above the first hand. The yield varies considerably from 26,000 to 55,000 kg per hectare.

Curing and Marketing

The ripening of banana is done in several ways, e.g. exposing the bunches to the sun, placing them over a hearth, wrapping them in closed godowns or smoking them in various ways. One of the common ways is to heap the fruits in a room and cover them with leaves, after which fire is lit in a corner and the room is closed and made as air-tight as possible. Ripening takes place usually in 30 to 48 hours. In a cool store, the bunches ripen well at about 15° to 20° C. The application of vaseline, a layer of clay or coal-tar to the cut-ends of the stalks prevents rotting during ripening and storage. Wrapping up the fruits and packing them in crates help to reduce the damage during transport.

MANGO

Mango (*Mangifera indica L.*) occupies nearly half of the total area under fruits in the country. It is adaptable to a wide range of soil and climatic conditions and grows well right from Assam to the southern-most limits of the country and from the sea-level up to about 1,500 metres. It withstands both fairly dry conditions and heavy rainfall, provided severe and recurring frosts in winter do not endanger the young trees.

Varieties

The number of varieties is very large. Each variety has its own peculiar taste, flavour and consistently of pulp. Some of the important commercial varieties grown in different regions are: 'Bombay yellow', 'Alphonso', 'Gopal Bhog', 'Zafran' (all early), 'Langra', 'Desheri', 'Safeda Lucknow', 'Safeda Malihabad', 'Fajrizafrani' (all mid-late). 'Fajri', 'Same Bihisht', 'Chausa', 'Taimura' (all late). In Uttar Pradesh; 'Bombai', 'Alphonso', 'Hemsagar', 'Krishna Bhog', 'Aman Dasheri', 'Gulab Khas' (all early), 'Langra', 'Aman Abbasi', 'Khasul-Khas' (all mid-late), ' Sinduri', 'Sukal', 'Taimuria' (all late) in Bihar; 'Bombai' or 'Maldah', 'Gopal Bhog', 'Hemsagar' (all early), 'Krishna Bhog', 'Zardalu' (both mid-late), 'Murshidabadi', 'Fazli Maldah' (both late) in West Bengal; 'Alphonso', 'Pairi', 'Cowsji Patel', 'Jamadar' in Bombay; 'Swarnarekha', 'Benishan', 'Cherukurasan', 'Panchadarkalasa', 'Desavathiyamamidi', 'Sannakulu', 'Nagulapalli', 'Irsala' in Circars; 'Rumani', 'Neelum Benishan', 'Bangalore', 'Alampur Benishan' in Rayalaseema; 'Murshidabadi', 'Mulgoa', 'Goabunder', 'Benishan', 'Neelam', 'Totapuri' or 'Bangalora' in Telengana; 'Alphonso', 'Peter', 'Rumani' in central districts; 'Mundappa', 'Neelam', 'Alphonso', 'Olour', 'Bennet Alphonso', 'Kalepad', 'Peter', 'Fernandin' in Coorg and Karnataka; and 'Padiri', 'Alphonso', 'Peter', 'Neelum', 'Bangalore', 'Rumani' in Tamil Nadu. In Goa, some excellent varieties like 'Alphonso', 'Fernandin', 'Mankurad' and 'Moussorate' are under cultivation. The new mango variety, 'Mallika' evolved at the Indian Agricultural Research Institute is now gaining popularity.

Propogation and Planting

Propogated vegetatively by inarching or budding in situ in the nursery, either by using Forkert or by using the T-method. The beginning of the monsoon in light-rainfall areas and the end of the monsoon in heavy-rainfall regions are the most suitable periods for inarching or budding. Recently, veneer-grafting has been found to be the best method of mango propagation. Grafted plant are ready for transplanting in the field after six to twelve months. Select straight-growing grafts and set them in pits filled with soil mixed with farmyard manure (45 kg) and a fertilizer mixture containing 0.225 kg of N, 0.45 kg of P and 0.225 kg of K per pit. The planting-distance is 7.5 to 9 metres in poor shallow soils and 15 to 17 metres in deep fertile soils. The beginning of the monsoon in low rainfall areas or the end of the monsoon in heavy rainfall tracts is the best time for planting. The graft-joint should be at least 15 cm above the ground.

Pruning

No systematic pruning is done. The removal of dead-wood and the thinning of over-crowded and mis-shapen branches after about four years is all that is necessary; flowers that appear during the first three or four years should be removed.

Culture

Before planting, the field is ploughed, harrowed and levelled. Thereafter, it is ploughed and harrowed twice a year, once in the beginning of the monsoon and again at the close of the rainy season or in the cold-weather. It is green-manured once every two or three years. Short-season intercrops, like vegetables, may be taken during the first four to five years. Young plants require irrigation regularly. After five to six years, when they have established themselves, the trees are able to grow and fruit satisfactorily without irrigation in most parts of Peninsular India. In northern India, they have to be irrigated throughout their life. Irrigation is usually withheld during the cold weather before flowering, especially in deep retentive soils. Though the exact manurial requirement is not known, regular manuring is beneficial. The dose recommended for the bearing trees is 45 to 70 kg of farmyard manure, 0.5 to 0.7 kg of N, 0.7 kg to 1.0 kg of P and 1.2 to 1.5 kg of K per tree. Nitrogen and half of potash may be given before the monsoon, and farmyard manure, phosphate and half of potash in October or before flowering starts.

Crop Irregularity

Grafted mango-trees bear fruits from the fourth or fifth year onwards and a full crop from the tenth or fifteenth year. The erratic bearing of mango is well known. It depends upon the variety , the weather and climatic conditions and cultural treatments. The selection of regular-bearing varieties, timely cultural practices and proper nutrition help to produce a regular crop. New growth in spring, on which flower-buds are produced during the next winter, can be encouraged by applying nitrogenous fertilizers (0.45 to 0.90 kg of N per tree). In the case of heavy late rains, an additional ploughing in winter helps to produce flower-buds in January-February. In the case of individual trees, ringing or girdling in August-September may also to help to force flower-buds the following winter. The application of Ethral (200 ppm) from September onwards has been found to induce flowering in mango in Karnataka by the Indian Institute of Horticultural Research.

Improvement of Old and Seedling-Trees

Mango-trees of inferior varieties, so also those raised from seeddlings, can be converted into choiçe varieties by grafting them in situ either by crown or side-grafting. In crown-grafting, the trunk of the tree is cut down to about half a metre from the ground and one or more scions of the selected variety are inserted into it between the bark and the wood by splitting open the bark. The scion should be a dormant, terminal shoot, about 12.5 mm in diameter, with a whorl of plump swollen buds at the top.

In side-grafting, the procedure is the same as in crown-grafting, except that the trunk of the stock tree above the grafting joint is cut down after the scions have sprouted and have established themselves properly. Old trees, having several branches, can be similarly improved by crown-grafting on each branch at a suitable height. Sometimes, the grafting is done by inarching, but the process is cumbersome, expensive and not very satisfactory.

Harvesting and Marketing

The fruit takes five to six months to mature. Depending upon the onset of flowering, the mature fruits are ready for harvesting from April to May in western India, from May to June in the Deccan, from February to March in Malabar, from April to July in the coastal Andhra Pradesh, from May to August in Mysore and Rayalaseema, and from June to August in northern India. The mature fruits are harvested by severing the stalks to which they are attached, when they are still green and hard. The signs of maturity vary with different varieties.

As a mango tree usually bears flowers in three or four distinct flushes lasting over a month, it is preferable to harvest the fruits as they mature. The fruits, so harvested, can be transported after packing them in baskets or wooden crates, properly padded with straw, wood-shavings or wool, to long distances. For overseas markets, they are packed in a single layer in specially designed wooden crates. For ripening, the fruits are spread out on rice straw in a single layer. Two or three such layers are built one above another in a well-ventilated room. The mangoes are ready for disposal after they change colour. Yield varies considerably with the variety, vigour of growth, flowering, etc. A grafted tree yields about 300 to 500 fruits in the tenth year, about 1,000 in the 15th year and 2,000 to 5,000 from the 20th year onwards.

CITRUS FRUITS

Citrus is grown in almost all the states of India. The total area covered is over 67,650 hectares, of which Madhya Pradesh, Madras and Maharashtra have the largest share. Citrus trees are grown in almost all kinds of soils, varying from heavy black soils to shallow open soils. Some of the varieties of citrus seem to adapt themselves to soil conditions better than others. They thrive in free-draining alluvial or medium black soil of loamy texture. A hard substratum or a sticky impervious layer is very injurious. Soils having a high water-table should be avoided. Though citrus trees on the whole do well in dry climate, with a rainfall between 75 and 125 cm, certain species, such as pummelo and certain mandarin oranges, thrive in heavy-rainfall areas of Konkan, Assam and Coorg.

GRAPEFRUIT

The name grapefruit has been derived from the habit of bearing the fruit in clusters like grapes. In India, its introduction is comparatively recent, and its cultivation is confined mostly to Punjab, the western parts of Uttar Pradesh and to places around Poona in Maharashtra.

Climate and Soil

The climatic and soil requirements of the grapefruit are similar to those of the orange. High rainfall and humidity are harmful, as they encourage diseases.

Varieties

The popular varieties, which are all imported, are 'Marsh Seedless', pink-fleshed 'Foster' and yellow-fleshed 'Duncan'.

Propogation and Planting

Propogation is done by budding. However, owing to their

polyembryonic nature, seedling trees have been frequently found to be quite satisfactory. The rootstock most successfully employed in the northern regions is kharna khatta. In the south and Bombay-Deccan, 'Jamburi' is commonly employed, whereas in Assam grapefruit does well on Rabab tenga. Planting, irrigation, manuring and interculture are the same as for the orange.

Harvesting and Marketing

The harvesting season is from January to March in the north and from September to November in the south. Picking, packing and other operations are the same as for the orange. Quality and flavour of the fruit is improved if it is stored before transporting.

LIME

The lime (*Citrus aurantiifolia Swingle*), both sour and sweet, known as kaghzi nimboo and mitha respectively, are more orized in India than lemon. Sour lime is propagated mainly from seed. Budding on rough lemon rootstock, layering and morcotting are also practiced to some extent. The tree is susceptible to frost. It flowers twice a year in February-March and again in August. The main crop is obtained in August from the first flowering. The second crop is ready in the following February.

The propagation of sweet lime is done from mature wood cuttings which root readily. It can also be propagated from seeds and the seedlings, usually come true to type. Planting of limes and their pruning, manuring, etc. are the same as for the sweet orange.

MANDARIN ORANGE

Climate and Soil

Santra or mandarin orange (*Citrus reticulata Balanco*) grows successfully in all tropical and subtropical parts of the country. It tolerates more humidity in summer and winter than the sweer orange. It is grown under rain-fed conditions in Coorg, Wynad tract, Palni Hills and the Nilgiris in the south between elevations of 600 and 1,500 metres.

In Assam, the main centres of production are the Khasi, Jaintia and Lushai Hills. The region around Nagpur produces a superior quality of mandarins. It is mainly grown under irrigation. In Punjab, its cultivation is confined mainly to the submontane districts up to about 600 metres. It can be grown successfully on a wide range of soils, but the ideal soil is medium or light loam with a slightly heavier subsoil. Heavy black soil, underlain with murram and having good drainage, is also suitable. In the Khasi Hills of Assam, oranges are grown on sandy or gravelly soils.

Varieties

The important varieties cultivated on a commercial scale are the 'Nagpur' orange, the 'Khasi' orange, the 'Coorg' orange, 'Desi Emperor' and the 'Sikkim' orange.

Propagation

The propagation of mandarin orange is largely through seed, except the 'Nagpur' and 'Emperor' varieties which are propagated by budding. Like other citrus species, the seed is polyembryonic. Therefore, while propagating by seed, the sexual seedlings which are usually stunted and poor are rogued out and the rest that are produced from the cells of the nucellus are allowed to grow.

The seedlings, thus selected, are more or less uniform in growth and production. They are, however, late in bearing and remain tall and slender. Budded plants do not suffer from these defects. The santra orange is usually budded on rough lemon (jambhiri, Soh-myndong or jatti khatti rootstock. The variety 'Emperor' is budded on the kharna khatta rootstock. Studies at the Citrus Experiment Station, Coorg, of the Indian Institute of Horticultural research, has shown that trifoliate, Rangpur lime, Kodakthuli and Troyer citrage are good rootstocks for mandarin.

Planting

In the hills and humid regions, where plantings are generally done on steep slopes, the land is properly terraced. In the plains, where the trees have to be irrigated, the land should be levelled. The trees are usually transplanted during the monsoon. In heavy-rainfall areas, the plantings is generally done at the end of the heavy rains. They are planted 4.5 to 6 metres apart.

Pruning

Prune young trees to build up a strong framework, as recommended for sweet orange. The bearing trees require little or no pruning. Undesirable growths, like water-shoots and crossing branches, should be removed once or twice a year. In Bombay-Deccan, root exposure or resting treatment is given to santra trees to make them flower to order.

Manuring

Farmyard manure, 20 to 25 kg per tree, is applied at planting, together with about half a kilo of ammonium sulphate. A mixture supplying 0.09 kg each of N, P and K per tree may be applied in the first year after planting, and the dose is gradually increased to 0.45 kg of each N and P and to 0.90 kg of K per tree in the seventh year and kept constant thereafter. The dose of farmyard manure is increased to 50 kg per tree. It may be replaced by green manuring. In northern India, manuring is generally done in winter, whereas in Bombay-Deccan it is done before the advent of the monsoon or at the time of root exposure.

Irrigation

When grown under irrigation, the method and frequency of application of water are the same as described under sweet orange.

Harvesting

Seedling trees bear their first crop in the eighth year and the full crop from the tenth year onwards. Budded trees start bearing from the fourth year and full crop is had from the seventh year onwards. The harvesting periods differ in different parts of the country. While picking the fruit, the stem-end should be cut close to the fruit without damaging rind. Packing is done by putting the fruits of different size grades in separate wooden crates.

SWEET ORANGE

Sweet orange is grown under both subtropical and tropical conditions. Dry and arid conditions, coupled with distinct summer and winter having low rainfall, are most favourable to the growth of the sweet orange. Rainfall seems to be unimportant if irrigation is provided, but atmospheric humidity exerts a great influence. The sweet orange can be grown on a wide range of soils, from heavy clays to very light sands, with pH ranging from 6.0 to 8.0. The tree is particularly sensitive to high concentrations of salts and cannot stand water-logging.

Varieties

The important varieties of sweet orange grown in each region are 'Blood Red', 'Pineapple', 'Hamlin', 'Jaffa' and Valencia Late' in northern India, mosambi in Western India and 'Sathgudi' and 'Batavian' orange in southern India.

Propagation and Planting

Usually, propagated by budding. The most suitable

rootstock is Jamberi or jatti khatti. For 'Pineapple', 'Hamlin', 'Jaffa' and 'Valencia Late' varieties, kharna khatta also provide a suitable rootstock. The trees are planted 6 to 7.5 metres apart each way in January or August-September in the north and in July-August in the south. The bud-union should be kept at least 15 cm above the ground while planting.

Pruning

The pruning of citrus-trees begins in the nursery. All branches that start within a few centimetres of the union are removed, leaving about a half a metre of clean straight stem with a few well-placed branches. All unwanted branches are removed once a month during the first year after planting, and once in two to three months in subsequent years. The bearing trees require little or no pruning. After the crop is picked, the branches touching the ground should be cut close total the laterals so that no stubs are formed. All diseased, injured and crossing branches, water-sprouts and dead wood should be removed periodically.

Root Exposure

In the Bombay-Deccan region, root exposure is given to the trees to bring them into flowering at a particular time of the year. Water is withheld for about two months in advance of the normal flowering season, and after about a month, the roots are exposed by removing about 10 cm of the soil in the case of light soils and about 20 cm in the case of heavy soils. After about 10 days, the soil is returned mixed with manure, and a light irrigation is given. After four or five days, a more copious watering is given, followed by 10 days later by the full dose of water. In the case of light soils, the withholding of water without root exposure is sufficient to check vegetative growth and force blossoming. In southern India, no root exposure is given nor is it feasible. In the north, the root-exposure treatment is not necessary as the trees normally rest in winter and flower once a year. It should be mentioned that in most situations, the root exposure of citrus trees is a devitalizing process and should be resorted to only under expert advice and direction.

Irrigation

After the first heavy irrigation given soon after planting a second light watering follows in four to five days. Thereafter, irrigation is given at regular intervals, depending upon the source of water and the nature of the soil. Under well irrigation, water is given after every eight days in hot months and about 12 to 15 days in cold months. Where irrigation is from canals, the usual interval is about 14 days. A light soil requires irrigation more often than a heavy soil. Excess watering should be avoided, especially in heavy soils. When the trees are young, irrigation water is applied in basins of about one-metre radius. The basins are enlarged as the trees advance in age. In the ring method of irrigation a bund is formed about three-fourths of a metre away from the trunk to prevent water from touching it. The furrow method ensures a more even distribution of moisture in the soil.

Harvesting

Trees begin to bear fruits from the fourth year onwards, but normal crops are borne from the seventh year. The main harvesting season in the north is December to February, whereas in the south, it is October to March. In the Bombay-Deccan region, there are two main seasons, November to January for ambe bahar, and March to May for mrig bahar Picking may be done any time during the day, taking care that the stem is cut close to the fruit without damaging the rind. The fruits are washed, dried and graded for size and packed into wooden cases for disposal.

GRAPE

Grape (*Vitis vinifera L.*) is a subtropical fruit which grows well in dry climates having a short sharp winter and a long dry summer. The vines shed their leaves and rest in winter, put forth new growth in spring and mature in summer. Grape does not thrive in regions having humid summers. It tolerates frost during the resting period, but succumbs to it readily during its growing period. In India, however, it grows under varying climatic conditions. In Punjab, Uttar Pradesh and Himachal Pradesh, it grows and fruits once a year in summer, and rests during winter. In southern India, where it is cultivation mainly in Maharashtra, Hyderabad-Deccan, parts of Karnataka and Tamil Nadu, the vine grows throughout the year and bears two crops, the first in April and the second in August-September.

The grape grows best on light, friable loamy soils with free drainage. Heavy soils are unsuitable.

Varieties

Varieties suitable for different regions of the country are:

1. *Northern plains*: 'Black Prince', 'Bedana', 'Foster's seedling', 'Kandhari Dakh' and 'Muscat of Alexandria' , 'Perlette'.
2. *Dry and temperate regions*: 'Thomson Seedless', 'Sultana' and 'Kishmish White'.
3. *Southern plains*: 'Bangalore Blue', 'Pachadraksha' and 'Anab-e-Shahi', 'Gulabi', 'Black Champa', 'Thompson Seedless'.
4. *Western plains*: 'Cheema Sahebi', 'Anab-e-Shahi', 'Thompson Seedless'.

Propagation and Planting

The vine is usually propagated by cuttings. In northern India, cuttings are made from one-year-old wood at the time of pruning in February, when the vines are dormant.

The cuttings are tied in bundles and stored in moist sand for about a month for callusing. The callused cuttings start well in the nursery. Elsewhere in India, they are obtained from the prunings in October and planted in nursery for rooting. The cuttings are ready for transplanting in January. It is also possible to raise a plantation by planting cuttings in situ in the field.

Planting

The rooted cuttings are transplanted in northern India in January and February when they are dormant. The planting-distance varies according to the method of training involved. It is 2.5m X 3.0m, if vines are trained on the head system and 6m X 6m, for the overhead pergola. In western India, it is 25m X 1m for the avenue system. in Karnataka, it is 4.5m X 6m and in Tamil Nadu 4.5m X 7.5m for the overhead arbour system.

Training and Pruning

The most popular systems of training are:

1. *Head system.* The vine is trained in the form of a dwarf bush. It is allowed to grow into an upright stem with the help of a support and the developing shoot is cut off at a height of 1 to 1.25 m in July. It is allowed to develop three to four lateral branches, each about 0.45 m long, arranged in the form of a vase or goblet. At the first dormant pruning in February, the lateral branches are shortened to spurs with one to two buds each. These spurs provide suitable arms for the framework during the succeeding summer. At the second dormant pruning, eight to ten arms, with one to two spurs on each arm, are retained for cropping in the third year. This system is cheap, but yields are poor. It is practiced in northern India on varieties which fruit on the first few buds of the cane.

2. *Cane system.* This system is suitable for training on a two-wire trellis. The trunk is headed back, as in the head system; four arms, two on each side of the trunk, are allowed to develop and are pruned to a 30-m length after a year. Fruit canes, each carrying 10 to 20 buds, develop on each arm which is tied to the trellis wire. A renewal spur is left on each arm just below the cane for the next year's crop.
3. *Cordon system.* After the vine has reached a height of 0.5 to 1 m, the stem is bent and is trained along the single-wire trellis. On each arm that develops from the trunk, short fruiting spurs, each carrying two to four buds, are allowed to remain at the time of pruning. The replacing spurs are also provided close to the base of the fruiting spurs.
4. *Pergola system.* In the pergola, arbour or bower system, the vine is allowed to develop into two or three branches, about 1 m from the ground. The branches are fastened to the horizontal wires of the pergola and allowed to grow and spread on the roof. The branches that grow on the arms are pruned each year according to the mode of bearing of the variety planted.

Pruning is usually done in northern India once a year in spring before the new growth starts. In Peninsular India, grapevine is pruned twice a year, once in summer and again in October, the exact period being decided by the distribution of rainfall. Sometimes, the girdling or ringing of a caneis carried out to hasten maturity and to improve the size and quality of berries.

Irrigation

The grapevine should be regularly irrigated. It is necessary to regulate the water-supply carefully both when the vine is in flower and when the bunches are ripening. Too wet a soil during those periods is not desirable.

Manuring

In addition to the farmyard manure (25 to 30 kg), a dose supplying 0.07 to 0.09 kg of N, 0.54 to 0.57 kg of P and 0.135 to 0.18 kg of K per vine at pruning is recommended. Green-manuring may be done whenever feasible.

Tips for successful grape-growing:

(1) The cuttings for planting should be selected from one-year-old dormant wood from healthy bearing vines.

(2) Pruning of the vines should be regulated to suit the bearing habit of each variety. The timing of pruning should be so regulated as to avoid the vines coming into blossom when the days are characterized by dewfall.

(3) A portion of the berries at the tip of each bunch may be thinned to ensure more effective spraying against diseases.

Harvesting and Marketing

The fruit is ready for picking after the berries near the tip have changed colour and have become sweet. The picked fruit should not be exposed to the sun and, if it is to be sent to a distant market, it should be packed in sawdust. Yields vary largely, depending upon the variety, the locality and upon a host of other factors. Yields even up to 40,000 kg per hectare have been obtained, though 15,000 kg may be considered an average. A single vine at Coimbatore has been known to yield more than 800 bunches in a single season.

GUAVA (*Psidium guajava* L.)

The total area under guava in the country is about 30,000 hectares, of which Uttar Pradesh has the largest area (9,840 hectares), followed closely by Bihar (4,800 hectares). It is a very hardy tree, withstanding heat and prolonged droughts, but is susceptible to frost. A cool winter induces heavy fruiting. It grows in all types of soils having pH ranging from 4.5 to 8.2.

Varieties

'Lucknow-49', 'Allahabad Safeda' and 'Seedless' are white-fleshed varieties. Several types having pink flesh and white flesh with bright red skin are also known.

Propagation and Planting

Guava is propagated through seed, and also vegetatively. Inarching, layering and air-layering are commonly practiced. Propagation through root suckers, root cuttings and budding is sometimes successful. Propagation is generally done during the rainy season. The new plants are ready to be set out after a year. The usual distance for planting is 5.5 to 6 metres.

Culture

The growing of a green-manure crop during the rainy season, and clean cultivation during the rest of the year are recommended. One or two irrigations between the end of the monsoon and the harvesting (winter) are given in northern India. In southern India irrigation throughout the year is necessary. In addition to bulky organic manures, the use of 45 to 60 kg of N, 77.5 kg to 90 kg of P and 100 kg to 110 kg of K per hectare is recommended.

Pruning

Young trees require pruning several times a year to prevent the formation of long and slender branches. As the fruit is borne on new growth, heavy pruning of the bearing trees increases fruiting. All flowers should be removed until the framework becomes strong enough.

Harvesting

Fruits must be plucked as they ripen. Plucking extends over

several weeks. For long-distance marketing, it is necessary to harvest the fruit somewhat earlier. Yields of 22,000 kg per hectare have been reported.

PAPAYA

Papaya (*Carica papaya L.*) occupies a very small area, yet its cultivation is widespread in the country. It grows well almost everywhere, except at altitudes higher than 1,500 metres. It cannot tolerate low temperatures. A dry warm climate is necessary. Strong winds are highly detrimental to the trees as the hollow stems break easily. Even though the tree is adapted to a wide range of soils, it grows best in the loamy soil. Deep clayey soils that are prone to water-logging should be avoided. In heavy-rainfall areas, a prolonged stagnation of water near its stem is highly injurious.

Varieties

'Washington', 'Honey Dew' (Madhubindu), 'Coorg Honey Dew', 'Singapore' and 'Ceylon' are important varieties. C.O.I., an improved strain of the Ranchi type has been evolved at Coimbatore. The varieties do not remain pure under the existing state of cultivation and give rise to varieties, both in tree and fruit characteristics.

Sex Variation

The papaya plant is normally unisexual. Some plants bear male flowers and some female. Occasionally, a plant with hermaphrodite flowers (having both male and female organs) may occur. The proportion of plants with male, female and hermaphrodite flowers varies with the variety. The proportion of fruit-bearing female plants of any variety varies from 40 to 60 per cent. The plants of 'Coorg Honey Dew' are either female or hermaphrodite and, hence, every plant yields fruits.

Propagation and Planting

Papaya is propagated through seeds. About 100 to 200 grammes of seed is required for an acre of plantation. Seeds obtained from mature large fruits, borne on female plants, are usually the best. The best time for raising seedlings is the monsoon period. The seedlings are ready for transplanting in four to six weeks, when they are 20 to 30 cm tall. They are lifted with a ball of earth around the roots; most of their leaves are clipped off, and they are planted in small holes made in the field 2.5 to 4 metres apart. Four seedlings may be planted in each hole about one-third metre apart from one another. After the plants have flowered, all male plants, except a few required for fertilizing the female plants, are pulled out. One male tree for every 10 to 20 female trees is sufficient. Transplanting is best done in the monsoon. The practice of planting four seedlings in one hole is not required for varieties, e.g. 'Coorg Honey Dew', which do not produce any male plants.

Culture

Papaya plants are irrigated once in every 10 to 12 days in winter and six to eight days in summer. To avoid stagnation of water near the trunks, the basin may be made to slope away from it. Nine kg of farmyard manure per pit is applied at the time of planting, followed by 35 to 45 kg of it every six months, once at the beginning of the monsoon and again in winter. A fertilizer mixture to supply 25 to 50 kg of N, 50 to 100 kg of P and 50 to 100 kg of K per hectare may be given in two equal doses every six months. The removal of weeds and a light or shallow ploughing or harrowing once or twice a year are necessary. Low-growing vegetables of short duration may be taken as intercrops. An occasional thinning of fruits is necessary to prevent overcrowding. Papaya may itself be grown as a 'filler' or in plantation of other crops where spacing is wide enough.

Harvesting and Marketing

Papaya flowers in about four months after planting and fruits are ready for harvesting in another six months. Except during winter, the trees continue to flower and fruit all the time. Fruits are picked when they are still hard and green, but show a distinct change in the colour of the rind. Yield varies from 30 to 150 fruits per tree. The packing of fruits in baskets in several layers should be avoided. A soft padding, such as of wood-shavings, wool or straw is recommended.

PINEAPPLE

Pineapple (*Ananas comusus (L.) Merr.*) occupies about 12,000 hectares and is grown mainly in Assam, West Bengal, Tripura, Uttar Pradesh, Andhra Pradesh, Kerala and Karnataka. It is a humid tropical plant and grows well, both in the plains and also at elevations not exceeding 900 metres. It tolerates neither very high temperatures nor frost. It grows in almost any type of soil, provided it is free-draining.

Varieties

'Kew', 'Queen' and 'Mauritius' are the three popular varieties. 'Kew' produces large fruits and is mostly used for canning. The other two have smaller fruits which are considered to be of superior quality. 'Kew' is a late-fruiting variety. 'Queen' is early, while 'Mauritius is intermediate.

Propagation and Planting

Pineapple is commonly propagated from suckers or slips. Suckers arising from the underground parts of the plant are commonly used. Slips arise from the fruiting stem and from the crown on top of the fruit. After the fruit is harvested, stalks are cut intc discs and used for propagation. Plants grown from suckers produce fruits in about 18 months, whereas those from slips and suckers propagated from disc cuttings take over two years. Suckers or slips are first cured by stripping off the lower leaves, followed by drying in the sun, or in partial shade for three to four days before planting. They are planted either in flat beds, where there is no danger of water stagnation, or in shallow trenches which are filled as the suckers grow and develop. Care should be taken to see that the bud or 'heart' of the suckers does not get buried. A planting density of 43,500 plants per hectare can be followed, keeping a distance of 30 cm between plant and plant, 60 cm between rows and 90 cm between beds. The rainy season is the best time for planting.

Culture

The field is prepared by ploughing, harrowing, etc., before planting. In the hills, proper terracing is a necessity. In dry regions, the crop requires regular irrigation. Even when rain-fed, irrigation during the dry weather is necessary every week or ten days. Twenty-five to 50 tonnes of farmyard manure per hectare is applied in two doses in the 6th and 12th month after planting. Sixteen grammes of N and 2.5 g of K_2O per plant should be applied. Of these, part of the nitrogen can be applied as foliar application as 4% urea. Only two suckers are retained on each plant for the ratoon crop. After harvest, the plants are earthed up to stimulate the rooting of the ratoon suckers. The plantations is allowed to remain on the same site for four to five years after which it is renewed.

Induction of Flowering

Uniform flowering can be obtained by the application of NNA (Planofix), calcium carbide or Ethrel (100 ppm) at the 45-leaf stage.

Harvesting and Marketing

Pineapple usually flowers from February to April and the

fruits are ready from July to September. Sometimes, off-season flowers appear, and they produce fruits in September-December. The fruits are harvested when they just begin to yellow and the eyes become full and the bracts wither. The fruit is cut clean, retaining with it about 5 cm of the stalk. The yield is 12 to 17 tonnes per hectare in the case of the two smaller varieties and 25 to 30 tonnes per hectare in the case of 'Kew' in the first year. With a high population density, even about 85 tonnes of fruit can be obtained. Fruiting decreased progressively in the case of the ratoon crops.

SAPOTA

Sapota (*Achras zapota L.*) known popularly as chiku in western and central India, is adaptable to a large variety of conditions of soil and climate. It flourishes in the heavy-rainfall areas of western and southern India and grows equally well in the drier parts of the Peninsular India. It is an evergreen, growing and flowering almost throughout the year. Rain or cloudy weather does not harm the setting of fruits. Young plants are easily injured by frost, but grown-up trees can withstand frost of a short duration. Its soil requirements are not very exacting, but it grows best in alluvial or sandy-loam soilshaving good drainage.

Varieties

'Cricket Ball' and 'Dwaropudi', both with round fruits, are common in Tamil Nadu, whereas 'Bangalora', 'Vavila Valasa' (both oval-fruited), 'Jonnavalasa' (round-fruited), 'Kirtabarati' and 'Pot' (dwarf trees) are recognized in Andhra Pradesh. In western India, 'Kalipatti' and 'Chatri' (both oval-fruited) are well-known varieties. The variety 'Cricket Ball' does very well in Karnataka.

Propagation and Planting

Propagated by layering, gootees (air-layering) or inarching.

Side-grafting and budding are also possible. Rootstocks employed are rayan or manikara (Manilkara hexandra or mohwa (Bassia latifolia and B. longifolia). The second and third species are not recommended, as they are not compatible with the sapota scions. In northern India, the planting-distance is 4.5 to 6 metres, whereas in the south, it is 9 to 12 metres. In northern India, the young plant must be suitably protected against frost and hot winds after planting.

Culture

Before planting, the field is ploughed, harrowed and levelled. Irrigation is given every six to twelve days, except during the rainy season. The removal of weeds and the loosening of the soil by ploughing or harrowing are done once or twice a year. Manures and fertilizers are applied, as recommended for mango. For the bearing trees, half of the dose is applied in October-November and the other half in February-March or before the monsoon. Intercrops of vegetables may be taken for the first six to ten years. No pruning is necessary.

Harvesting and Marketing

Substantial fruiting starts from the fourth or fifth year. The fruit takes four months to mature. Flowers appear throughout the year, but the crop is available for harvesting in two to three seasons, i.e. northern and central India, March-April and August-September; in southern India, February-June and September-October and in western India, January-February and May-June. The yield varies from 200 to 300 fruits in the fourth year, from 700 to 800 in the seventh year, from 1,500 to 2,000 between the tenth and 15th year, and from 2,500 to 3,000 from the 20th to 30th year. Mature fruits show a yellow streak, whereas the immature fruits show a green streak when scratched with the fingernail.

POMEGRANATE

The largest area under pomegranate (*Punica granatum L.*) about 500 hectares, is in Maharashtra, but there are small plantings in almost all parts. Where winters are cold, the tree is deciduous, but in the plains, it is evergreen. A hot, dry summer produces the best fruits. The presence of many seeds and of tannin in the rind and membranes detracts from its attractiveness. The tree can stand considerable drought.

Climate

Pomegranate is sub-tropical fruit. It can adopt itself to a wide range of climatic conditions and can grow up to 1800 m above sea level. The fruit tree grows best in semi-arid climate where cool winter and hot and dry summer prevail. The tree requires hot and dry climate during the peroid of fruit development and ripening. The optimum temperature for fruit development is 38°C.

The tree can not produce sweet fruits unless the temperature is high for a sufficient long period. Under tropical and sub-tropical climate, it behaves as an evergreen or partially deciduous. Under humid conditions, the sweetness of fruit is adversely affected. Therefore, it is considered that pomegranate is a hardy fruit and can thrive well under drought conditions, though yield is low. The plant bear well only under irrigation. In areas of low temperature, the tree behaves as deciduous in nature and sheds its leaves during winter months. It is also rated as winter hardy fruit tree. Two years old shoots of hardy cultivers are not damaged even at 2°C. When temperature falls below -14°C, visible damage due to frost occurs.

Soils

It can be grown on diverse types of soil. The pomegranate is not very particular about its soil requirement. The deep loamy and alluvial soils are ideal for its cultivation. It can tolerate soils which are lomay and slightly alkaline. It can thrive well on comparatively poor soils where other fruits fail to grow. Pomegranate can also be grown in medium and black soils. It is rated as salt-hardy fruit plant.

Choice of Varieties

1. *Ganesh*: It is an improved variety known as GBG No. 1. Ganesh is a seedling selection by Dr. G.S. Cheema at Pune. It is a selection from Alandi and considered to be the best variety. The fruit is medium in size. It has soft seeds. Ganesh is a high yielding variety and is a good cropper. The flesh is pinkish and has juice with agreeable taste.
2. *Alandi*: Fruit medium in size, fleshy tasta, blood red or deep pink with sweet slightly acidic juice. Seeds very hard. The variety is named after the name of village where it was grown extensively.
3. *Dholka*: Fruit large size, rind grenish white, fleshy testa, pinkish white or whitish with sweet juice. Seeds soft. Juice is acidic. It is medium cropper.
4. *Kandhari*: It produces large fruits. The rind is deep red. The flesh is dark red or deep pink. The juice is slightly acidic. The seeds are hard. The variety is successfully grown in Himachal Pradesh.
5. *Muskat*: This variety is also largely grown. Fruits small to medium in size. Rind in somewhat thick. Fleshy testa, with moderately sweet juice.
6. *Nabha*: Tree is medium in growth, fruit skin yellowish and smooth. Cracking medium. Juice is 65 per cent. The arils are light pink in colour. Seeds are medium in soft.

Propagation

Pomegranate plants raised from seed vary widely and are undesirable. Thus, they must be raised vegetatively.

Among the vegetative methods of propagation, cuttings are universally used for raising pomegranate plants on commercial basis.

Mature wood is used for making the cuttings and these are 8 to 10 cm long. The cuttings are planted in the nursery fields in such a way that not more than one-third of the cutting is exposed. The best time of making the cuttings is December-January when the plants shed leaves. The cuttings made during September-October can also root satisfactorily. The cuttings in the nursery field are planted directly after making them from the plants. They need not to be kept for callusing.

Pomegranate may also be propagated by air-layering or gootee. Treatment with 10000 ppm Indole Butyric acid in lanolin as carrier was found to improve rooting. Ground layering is another method used for multiplying the pomegranate plants.

Planting Operation

Land is prepared thoroughly and levelled prior to pit digging. The layout is done following square or hexagonal system. The size of the pit should be 60 × 60 × 60 cm. Pits should be filled with 22 - 25 kg of farmyard manure or compost, 1kg of superphosphate and good soil mixture.

The plants are planted at a distance of 6m × 6m apart in square system and it will accommodate 275 plants per hectare. In higher but deeper soil, the planting distance can be reduced to 5 × 5 m.

Planting density is the most important yield contributing factor which can be manipulated to attain the maximum production per unit area. The optimum spacing is important for the maximum utilization of land and good income over a long period. At MPAU, Rahuri, it was observed that as the plant density was increased, yield per hectare also increased without affecting fruit quality. A density of 1000 plants per hectare gave 2-3 times higher yield and 2.44 times more profit as compared with normal plant population of 400 per hectare. It was also recommended that for higher yields for the first four to five years after planting, a distance of 5 × 2 m may be adopted and alternate plants may be removed afterwards maintaining a planting distance of 5 × 4 m. The best time of planting pomegranate in Northen India is dormant period, i.e. January to mid February and in South India during monsoon season.

Flowering and Fruiting

In evergreen pomegranate cultivers, the flower buds of the spring flush are borne on nature wood of one-year old shoot, whereas the flowers which appears during July-August are borne on the current year's growth. In deciduous cultivers, the flowers are borne on the current year's growth between July and August. The flowers are found mostly in clusters, either terminally or in axils of the leaves. The inflorescence in cyme and due to heavy drops of secondary and tertiary buds they appear to be solitary in clusters. In Western India, three flowering seasons and in North India, two flowering seasons have been reported. The flowering period of different cultivers is also quite variable. Under Delhi condition, Dholka, Kandhari, Muskat and Patiala flowered only once in a year while Ganesh and Japanese Dwarf flowered twice.

Orchard Cultural Practices

Irrigation

The newly set plants require regular irrigation so that the roots become well established and the plants can start growth. The plants may be individually watered daily or about a week after planting. In northern India where planting is done during the spring, regular watering may be given every 7 to 20 days till the start of the monsoon. In areas where planting is done during the monsoon, irrigation may be given whenever there is no rain for a prolonged period of time. After the plants are well established, in about 6 months, they can stand considerable amount of drought and irrigation may be given at intervals of 2 to 4 weeks depending upon the soil, climate, weather conditions and intercrops grown. Regular irrigation is essential from flowering to ripening of fruits, as irregular moisture condition results in dropping of flowers and small fruits.

Intercropping

Intercropping is pomegranate orchard is highly desirable because it takes about 6 - 7 years to come to commercial bearing. Vegetables viz. cabbage, cauliflower, tomato, radish, cucurbits, moong, peas, beans or green manure crops can easily be followed in pomegranate orchard. The growing of intercrops should be carried out for the first four years of the life of the plantation. It is usually advisable to allow the intercrops to grow throughtout the year. Intercrops can be continued for another 3 to 4 years after the plants had started bearing. It is best to grow a green manure crop during the monsoon and burry, when it has completed its vegetative phase and started flowering.

Manuring and Fertilization

In northern India, manures are applied during February, whereas in other areas, manuring may be done just before the start of monsoon in case of young plants. The one year old tree should be manured with about 10 kg of farmyard manure and 150 to 200 g of ammonium sulphate. The amount is increased by the same amount every year so that a five year old tree gets 50 kg of farmyard manure and one kg of ammonium sulphate.

At Mahatma Phule Krishi Vidyapeeth, Rahuri, an application of 500 g N, 125 g P_2O_5 and 125 g K_2O is recommended for 'Ganesh' variety. Under Udaipur conditions of Rajasthan, application of 240 g N, 160 g P_2O_5 and 60 g K_2O per plant is suggested for variety 'Jodhpur Red'. In Gujrat state pomegranate is applied with 50 kg farmyard manure, 500 g N, 250 g P_2O_5 and 500 g K_2O per plant per year. Application of fertilizers is done in December-January for Ambe bahar, in May-June of Mrig bahar and in October-November for Hasth bahar.

In Andhra Pradesh, adult bearing trees are applied with 800 g N, 400 g P_2O_5 and 400 g K_2O per tree alongwith 100 kg FYM per tree. In Karnataka a dose of 200 kg N, 300 kg P_2O_5 and 100 g K_2O alongwith 12.5 tonnes per hectare farmyard manure is applied. In Orissa, 45 kg N, 115 kg P_2O_5 per hectare along with 10 kg FYM + 100 g sterameal per tree are applied. In Tamil Nadu, 600 g N, 500 g P_2O_5 and 1200 g K_2O alongwith 30 kg FYM is incorporated per tree per year.

Training and Pruning

Training: Pomegranate may be trained as multi-stemmed tree or single stemmed tree.

(i) *Multi-stemmed tree*: In this method 3 - 4 stems are left at hill and remaining shoots are removed. In Maharashtra, the growers prefer multistem training by retaining all stems. But yield has not been found to be affected by number of stems per plant.

(ii) *Single-stemmed tree*: The single stem is left by removing all the side shoots at the time of planting. The main stem is headed back at a height of about one metre results in the formation of branches. Four or five well distributed branches on all sides above 60 - 70 cm from the ground level are allowed to grow. In the third year of planting one can maintain desired shape of the pomegranate. Single stemmed tree has tendency to produce less number of shoots.

Pruning: Pomegranate does not usually require pruning except for removal of suckers, dead and diseased branches and developing a sound framework of the tree. It is essential to remove the suckers as soon as they arise. The fruits are borne terminally on short spurs produced all along the slow growing mature wood. These bear fruits for 3 to 4 years. Therefore, only a limited pruning of bearing tree is required. Annual pruning in winter during dormant period should be confined to shortening of the previous season's growth to encourage fruiting.

For getting a good crop, a set of new shoots should be allowed to develop every year on all sides of the tree and gradual growth of new shoots should be encouraged by restricted cutting back of the bearing shoots.

Crop Regulation

The pomegranate starts fruiting about 4 years after planting and continues for about 25 to 30 years. Economic yield is generally obtained after 10 years of planting.

To regulate flowering, water is withheld for about two months in advance of the normal flowering season. After two months, manures and fertilizers are applied and light irrigation is given. Three to four days later, heavier irrigations at normal interval are followed. The tree readily responds to this treatment by producing new growth and blooms and bears a good crop.

A full grown pomegranate has tendency to bear flowers and fruits throughtout the year. To obtain higher fruit yield during a particular period, plants are given a resting period by which the natural tendency of the tree is altered with artificial means. It is done by withholding of water for about 2 months in advance of normal flowering, root exposure and also use of chemicals. By adopting such methods flowering can be induced in June-July (Mrig bahar) coinciding with the break of monsoon, February-March (Ambe bahar) and September-October (Hasth bahar).

Mrig bahar is taken in Deccan areas where water is so scare during the hot weather. The flowering, therefore, is so forced that the maximum requirement of water falls during the rainy season. For this treatment, watering is withheld from December to April-May results in sufficient suppression of growth. In the month of March-April leaves are shed as plants go in dormant stage. The manures and fertilizers are applied and light irrigation is given which is followed by two heavy irrigations at 7 days interval before rain sets in. Within 15 days, trees will put on profuse growth along with the formation of flowers and fruits. The fruits ripen in October and continues upto December.

Ambe bahar is taken in the areas where enough water is available during hot weather. The fruits are available during June and July and no irrigation is given after the start of the rainy season. The trees shed their leaves by October-November, when a shallow hand digging or ploughing is done. During December-January, manures are applied. The first irrigation is given in January and the flowers appear within a month of this irrigation. In dry regions of western Maharashtra Ambe bahar has been found to be better treatment than Mrig bahar. Hasth bahar is seldom taken. The trees have to be made dormant during August-September. This is rather uncertain because of the rains that occur during this period.

Pollination

Both self and cross pollination are recorded in pomegranate. Greater percentage of fruit set was observed by hand pollination and pollination under natural conditions i.e. open pollination.

Harvesting and Fruit Handling

Pomegranate is a non-climacteric fruit. Its fruit become ready for harvesting in 5 - 7 months after the appearance of blossoms. Mature fruits become slightly yellowish and further pink to red. On tapping, the fruits give metallic sound and when pressed they give a 'Crunch' sound. The fruits are harvested with the help of secateurs.

The trees begin to bear fruit in the fourth year when a small crop of 20 - 25 fruits (4 - 5 kg) per tree may be harvested. In the 10th year, it rises to 100 - 150 fruits (20 - 25 kg). The average yield in well managed plantation may be as much as 200 - 250 fruits per tree. After harvesting, sorting of fruits should be exercised to remove undesirable fruits. The healthy fruits are packed mostly in bamboo baskets and wooden crates containing 10 - 12 kg with a padding of paddy straw or dry grasses. The fruits can be transported to distant market without any loss.

Cracking or Splitting of Fruits

This disorder is reported to be due to boron and calcium deficiency. There is further attack of insects or fungal attack on the cracked fruits. So fruits become unfit for marketing. The Mrig bahar crop is more susceptible to cracking than the crop of other bahars.

Main cause of this malady is the wide variation in moisture content of the soil as well as in the humidity of air due to monsoon. If there is sudden break in the rains during August, the growth of fruit is arrested. So as a result of the dry atmosphere that follows, the elasticity of the skin is lost; then there is rain again and growth restarts, results in cracking of fruit skin.

Thus cracking fruits is mostly due to irregular water supply to the trees. Ambe bahar crop is regularly irrigated, so it does not crack badly. The best treatment is to give regular irrigations to the Ambe-bahar crop taking care that at no stage there is a scarcity of water. In case of Mrig-bahar crop, the splitting of fruits cannot be controlled altogether as the variation in humidity cannot control cracking and can, however, be minimised if the plants are regularly irrigated whenever there is a break in rain.

Insect Pests

Fruit borer or pomegranate butterfly

This is a serious pest found all over India. Infestation starts from flowering to button stage. The caterpillar bore inside the developing fruits. Such infested fruits are also invaded by bacteria and fungi and cause fruit to rot. Such affected fruits fall down. Collect and destroy the affected fruits. Apply carbaryl 0.2 % @ 4 g/litre or phosphamidon @ 0.3 ml/litre of water at 10 - 15 days interval. Also, bag young fruits with coarse cloth or muslin cloth or polythene of 300 gauge thickness.

Bark eating caterpillar

This pest bore into the bark of pomegranate tree and feed inside. Trees become weak and do not bear fruits.

Avoid over crowding of trees by removing unwanted twigs. Clean the affected portions by removing all web. Inject kerosene oil or petrol and plug the hole with cotton wool soaked in carbon bisulphide.

Stemborer

The caterpillar of this pest makes a hole and bores through the main trunk or main branches. It comes out at night and feeds on bark. Clean the hole by removing insect excreta with the help of a hooked wire. Plug the hole with cotton plug dipped in petrol, chloroform, carbon bisulphide or kerosene oil followed by sealing it with mud or painting with coaltar.

Sap sucking insects

These are mealy bugs, scale insects, thrips, aphids, mites cause shedding of buds, flowers and fruits at very young stage.

(i) Spray 0.04 per cent Monocrotophous for the control of mealy bug and scale insects.

(ii) Spray 0.04 per cent Dimethoate or phosphamidon for the control of white flies, aphids and thrips.

(iii) Spray water soluble sulpher 1.25 g/litre for the control of red mites.

Diseases

Leaf spot

The former pathogen is bacterium and the latter is fungal. The bacterium pathogen causes irregular water-soaked spots on leaves. They are light brown to dark brown in appearance. The leaf spot caused by fungus, produce minute violet black or black spots on leaf. For their control, collect fallen leaves and fruits and destroy them. Spray 0.2% Captan or Dithane M-45 at fortnightly interval.

Fruit rot

The flowers are affected and fail to set fruit. The young fruits may drop pre-maturely. Yellow or black spots appear all over the fruit. The disease spreads through the seeds of affected fruits. The incidence is wide spread during rainy season. Remove all affected twigs, fruits and burn them. Spray Dithane Z-78 at 0.2 per cent at fortnightly interval.

AVOCADO

The avocado (*Persea sp.*) fruit is rich in protein and fat, both of which are deficient in the average Indian diet. It is neither sweet nor juicy, and is eaten fresh, either with bread or in salads. Of the three recognized races, the Mexican seems unsuited to India, the Guatemalan succeeds best at elevations above 1,000 metres, and the West Indian succeeds in the plains, except in the drier areas. Large parts of Maharashtra and southern India are quite suitable. A well-drained soil is required.

Varieties

About a dozen varieties are grown in different states. 'Pollock', 'Paradeniya Purple Hybrid' and 'Feurte' were introduced from Ceylon and have established themselves successfully at the foot of the Nilgiris.

Propagation and Planting

The easiest method of propagation commonly employed in India is through seeds. In Maharashtra and the Nilgiris, layering has met with success. The plants are set out 7 to 9 metres apart.

Harvesting

The trees generally flower between February and April and the harvesting of the fruits is done in August-September.

CASHEW

Cashew is grown both for its fruit as well as for its nuts, mainly for the latter. The crop is grown chiefly in Peninsular India, particularly along the coast. There are very few regular plantations of this crop. Cashew cannot tolerate the severe summers or winters of northern India. Even in the south, it does not grow satisfactorily at elevations higher than 300 metres. It is not very exacting in soil requirements, as it grows even in very gravelly soils. However, it needs a free-draining soil. It grows in areas having rainfall varying from 50 to 400 cm. Continued adequate soil moisture is, however, necessary for the success of cashew plantations.

Varieties

There are no distinct varieties of cashew and it exhibits a marked variation in fruit and nut characters when grown from seed. Some of the superior variants can be multiplied or perpetuated through vegetative propagation.

Propagation and planting

Sowing the seeds in situ is the usual practice, but it is possible to transplant one-month-old seedlings after cutting them back by about one-third. This fruit can also be propagated by air-layering, inarching and side-grafting. The planing distance varies from 6 metres in lateritic and rocky soils to 12 metres in deep loamy soils.

After-care

No attention is paid to cultivation, irrigation or manuring. An occasional clearing of undergrowth and the pruning of dead and diseased branches is necessary to maintain the trees in health and vigour.

Harvesing

The fruits ripen from March to May but the season is prolonged during the years when heavy rainfall is experienced in November-December. The first bearing is normally secured in about three years after planting, though satisfactory crops can only be gathered after about eight years. The yield of nuts varies from 110 to 220 kg per hectare.

Curing and Processing

The nuts are separated from the cashew apples immediately after harvest. The dried nuts are roasted either in open pans over a furnace or in rotary cylinders and in oil-baths. Shelling is done by hand soon after roasting. The kernels, so obtained, are dried in the sun or in hot-air chambers. They are then kept in sweating chambers for some time. The nuts are then ready for grading and packing. For the export trade, the kernels are packed in tins under vacuum or in carbon dioxide. For internal markets, the kernels are sometimes packed in tins of different sizes and they may or may not be hermetically sealed.

JACKFRUIT

Jackfruit (*Artocarpus heterophyllus Lamk.*) is grown in southern India as stray trees in home gardens and coffee plantations where it flourishes in the humid climate on hill slopes. Warmer plains are suitable, provided there is adequate soil moisture. Cold and frost are harmful.

Propagation and planting

Usually, seedlings are planted. Inarching is employed to propagate jack trees vegetatively. A spacing of about 12 metres is necessary for planting in the fields.

Harvesting

The 'Singapore' variety produces fruits in 18 to 36 months after planting. Others take about eight years to come into bearings. The fruits are in season from March to June, the season extending up to September at higher elevations. Yields range from a few to 250 fruits per tree, each fruit weighing from 9 kg to even up to 20 kg.

LOQUAT

Loquat (*Eriobotrya japonica Lindl.*) is grown mainly in Punjab and Uttar Pradesh on about 800 hectares each, and to a small extent in Delhi, Assam, Maharashtra and in the hills of southern India. Whereas the tree can stand temperatures well below freezing, the flowers and young fruits, which are borne in winter, are severely injured by frost. It does not fruit well in areas with warm winter. It thrives on many different types of soils.

Climate

The loquat has been naturalized in India. It is grown nearly throughout the country upto an elevation of 1525 m above sea level. It can be grown throughout the tropics where there are elevations of a few thousand feet. The loquat succeeds well under subtropical climate. It can grow successfully in regions in which the temperature does not fall below the freezing point. In general, a mild climate with an average annual rainfall of 60 - 100 mm, well distributed throughout the year, is ideal for loquat cultivation. Since the tree bloom between November and late January at certain places, the crop may be destroyed by moderate winter frosts. The fruit of loquat is most susceptible to frost injury when it just starts colouration. Heavy damage is noticed in early ripening varieties.

In the areas where scorching hot winds begin to blow before the fruit ripens, the fruit either remain too small or do not ripen properly. Under such conditions, the pulp does not produce abundant required quantity of juice. Further, the fruit become sunburnt and unfit for marketing. Under cool and foggy weather at the time of ripening, the fruits lack in sweetness and flavour. The loquat trees are, however, resistant to heat and drought. The thick, leathery leaves are well adapted to withstand seasons of neglect without serious injury. Althought an evergreen fruit, loquat is unique in tolerating cold weather conditions. However, warm and dry climate is essential at the time of fruit ripening. In arid and dry zones, the fruits are prone to sun burn injury. Consequently, loquat should be cultivated only in the submontane or other areas with mid climate or places free from severe hot weather condition.

Soil

The loquat can thrive in a wide variety of soils. In some places seems to do best on a light sandy loam; whereas at other places it has faired well on heavier soils. Good drainage is, however essential and the subsoil should be free from hard pan. Stagnation of water in the land, even for a short time may damage or kill even grown up trees and such soils should be discarded. According to the opinion of an expert from Israel, heavy soil of medium type should be preferred than a very light soil.

Choice of Varieties

Golden Yellow: It has medium sized, egg-shaped fruits with attractive golden yellow colour. Flesh is yellowish which tastes sour-sweet. Each fruit contain 4 - 5 dark brown, medium sized seeds.

Pale Yellow: It has large fruits, which are slightly conical to roundish in shape and pale yellow in colour. Flesh is white and tastes sour-sweet. Each fruit contains 2 or 3 medium-sized seeds.

California Advance: Fruit medium-sized, conical to round in shape, external colour yellow. Flesh creamy white, sour-sweet in taste. Fruit contains 2 or 3 medium-sized seeds.

Tanaka: Fruit mediun, 4 - 5 cm long, 3 - 7 cm broad ovate in shape, smooth and moderately pubescent. Rind medium, firm. Pulp medium completely filled, sayal brown, coarse firm, juiceness abundant, taste pleasant subacid. Seeds 2 to 4 per fruit, medium in size.

Propogation and Rootstock

The loquat seeds germinate readily when it is sown immediately after extraction from the fruit. The seed should not be allowed to dry after extraction as exposure to heat and light tends to result in poor germinaiton and stunted seedling growth. Fresh seed are sown during April-May in moist sand for germinaton. When the seedlings are 4 - 5 cm tall, they are transplanted in the nursery under the mother trees for inarching. If the mother plants are high headed, the seedlings are transplanted in the pots and brought in contact with the mother plants by raising platforms when they attain inarchable size. The seedlings grow rapidally and are fit for grafting in the following rainy season.

Promising loquat varieties are generally grafted upon domestic loquat seedlings of the commercial varieties. Several other rootstocks such as pear, apple, quince (*Cydonia oblonga*) and Mespilus have also been used in certain occasions.

Propagation Techniques

The usual technique followed is to raise the plants through inarching. The best time for inarching is July-August. Air-layering can also be tried but success is much less. In air-layering 3 months old shoots are used. The use of 3 per cent indole butyric acid enhances the success in air-layering.

Planting Operation

There are two planting seasons for loquat plants viz.February-March and August-September. The plants should preferably be planted during August-September when the weather has cooled down sufficiently. In general, loquat is planted at a distance of 6.5 m × 6.5 m in square system, thus accomodates 225 plants per hectare.

The pit should be made 1 × 1 × 1 m. While mixing add 3 - 4 baskets of farmyard manure and 200 g of single superphosphate per pit. Add 30 - 50 g BHC 10% dust to ward the attack of white ants.

Flowering and Fruiting

In India, the flowering period of loquat is very long, lasting from mid July to January or sometimes even upto May. Three reproductive flushes under tropical conditions are noted, out of which the intermediate ones give the higher yield of better-sized fruits. In Saharanpur only one flesh continued flower from September to February. The number of inflorescence was higher in begining but only a few of them bore fruits. The number of inflorescence continued to decrease with the advancement of season, but the percentage of fruit-bearing panicles gradually decreased. The fruit size seemed to the inversely proportional to the number of fruits per panicle. Earlier panicles gave best-sized fruit and size reduced afterward.

In northen India only one continuous reproductive flush appears. No peak hour of anthesis was observed. The opening of flowers continued throughout the day. The dehiscence of anthers took place in longitudinal fashion.It took more than a day to complete dehiscence in all the varieities except improved 'Golden Yellow' and 'Pale Yellow' in which case it was completed only in one day. The time of dehiscence was found to have hastened with the increase in temperature.

In Punjab, the flowering period in loquat is very lengthy. The flowering starts sometimes in the first week of October and continues upto third week of December. The number of flowers per cluster may vary from 50 to 100 but in general, not more than 15 - 20 fruits per cluster are set. Some of the varieties are good pollinizers for others.

Loquat trees grow singly or in small groves, though produce perfect flowers, yield negligible or no crop. This is reported to be due to self-incompatibility in commercial loquat varieties. The edible portion in loquat is entirely toral in nature, consisting of pith and cortical areas. The development of edible portion consists of rather uniform growth of receptacle tissue throughout the fruit. The toral cells of mature fruit are large, thin walled and very juicy.

Orchard Cultural Practices

Irrigation

The loquat is more drought resistant than any of the citrus fruits. However, the best results are obtained when the orchard is irrigated judiciously. There must be sufficient moisture in the soil in order to enable the shoots to develop and the mature terminal buds to fill out properly. As the trees blossom buds. During fruit growth to maturity three to four irrigations are generally advisable.

Interculture

Through cultivation of loquat field should be given to check the weed growth. It prefer clean cultivation management practice. Mulching with brown, black, or transparent polythene film from November to June in loquat orchard was found effective. The cultivation of leguminous over crops is, however, considered beneficial. Winter cover crops may be planted before September, the purpose is that their sufficient growth to be turned under before the harvesting starts. The leguminous crops like gram, peas, mash, etc. should be preferred as intercrops.

Manuring and Fertilization

Although little work has been done, it is generally recognized that loquat tends to exhaust the soil and that for good regular cropping it needs adequate nutrition. The farmyard manure should be applied in September alongwith entire quantities of phosphorus and potash. However, one-half dose of nitrogenous fertilizer should be applied in October before flowering and remaining half in February-March i.e. after the fruit-set.

Training and Purning

Loquat is generally trained according to Central leader or open centre system. During initial stages, stem up to height of 50 cm should be kept clean. Later on, purning is confined to removal of dead and diseased branches in the full grown tree. Harvesting of mature bunch is a kind of purning and it encourages new growth. The best time for purning is during summer after the crop has been harvested.

Top-working

A large percentage of the loquat seedlings trees are unprofitable. Such healthy trees should be top-worked with some improved commercial varieties. The vegetative method of propagation such as inarching, T-budding and bark grafting could be employed. Grafting plants will start bearing superior quality of fruit after third years. Top-working of inferior loquat tree should be done during May. Good number of healthy shoots will emerge from the headed back plant during the rainy season. Only one or two healthy branches should be retained for grafting purpose.

Thinning of Fruits

The loquat is a prolific bearer fruit plant. The tree has tendency to over bear. Thinning of fruit in the clusters or removing of some of the clusters may, therefore, be practised, for improving the size and quality of the remaining fruits. The thinning should be done when the fruits are less than 1.2 cm in diameter.

Harvesting

The loquat tree starts bearing fruit after third year of planting. The yield goes on increasing as the trees grow older and maximum yield is obtained after fifteen years of age. The harvesting of immature and green fruits should be avoided. All fruits in a cluster usually mature uniformly, the entire cluster may be cut at once. But, in some cases where the fruit at the base ripens before that at the tip of the cluster, the pickers will have to clip the ripe fruits by hand. The fruits should never be pulled from the tree by hand, as the stem would separate from the flesh and cause dacay to set in at once. The fruit should preferably be harvested with clipper.

The average yield of loquat tree is about 16 kg. Well maintained and healthy tree can yield fruit upto 40 kg.

Grading

The fruits should be suitably graded before packing into two grades. The large sized fruits tree from blemishes should be placed in one box while all the remaining marketable fruits should be packed separately. The undersized and mis-shapen fruits should be collected in a separate lot which can be sold for the preparation of jams, jelly or other products. All superfluous stem should be clipped off and badly bruised, shrivelled or scarred fruits should be discarded.

Packing

The fruit requires careful packing. Paper is placed at the bottom of each box. The large sized choice fruit should receive better handling and packing. The paper cuttings can be used for providing cushion. The wood boxes of 14 kg size should be used for sending fruits to nearby market. However, for distant markets, smaller packages are used to give considerable protection to the fruit.

Diseases

1. Shoot/fruit blight and Bark canker

The disease is caused due to fungus Phoma glumerata. The cankers appear on the bud scars, wounds, twig stubs or in crotches. Small circular brown spots appear around a leaf scar. As the canker enlarges the centers become sunken with the edges raised above the surrounding healthy bark. The fungus perpetuates it self on the trees in bark cankers.

The canker should be removed and the dead bark decorticated alongwith 2 cm of healthy bark. The dead-wood and purnings should be destroyed. The wounds should be covered with a disinfectant solution and painted with Bordeaux mixture immediately afterward and also in March and June.

Crown rot

It is caused due to fungus Phytophthora sp. The water loving fungus attacks the bark producing canker extending from the ground level up to the points where the main stem bifurcates. The rot girdles the trunk during the next 2 - 3 seasons. Flowering is very profuse on the affected trees, but fruiting is sparse and of low grade. Yellowish green foliage is characteristic of the crown rot disease. Often the half side of the tree affected by crown-rot may show symptoms. Some branches may be killed every year and ultimately the whole tree may succumb to the disease and dry up completely. The pathogen is soil borne and perpetuates itself in the dead cankers.

Remove the severely infested trees and use them as fire wood. Remove the diseased bark during the dry season by extending the cut an inch beyond the diseased zone on all sides. First treat the cut with a disinfectant solution and then apply Bordeaux paint after a week. Immediately after this, spray the tree with 2: 2: 250 Bordeaux mixture. Repeat the spray just before the mansoon, during the mansoon and thereafter, too, till October.

Root-rot - White-rot

The disease is caused due to Polyporus palustris. The bark and the wood of the root including the root collar is affected. The decayed wood is pinkish to dull violet in colour whereas in the advanced stages, small, white, elongated pockets appear and they form a mass of spongy white fibres. The affected tree begins to show symptoms of wilt, early leaf fall and increase in the fruit-set. The fruiting bodies conks (gidder peehree) which may grow up to 30 cm or more in diameter usually appear when the rot is fairly well advanced. They are either hidden by the litter or lie exposed on the surface of the soil. Locate the affected trees in early stages by examining the roots and the root-collar region of the tree showing weakening signs. Digout decayed roots and cut them completely right from the collar region. Treat the cut end of the roots immediately with disinfectant solution. When dry, apply Bordeaux paste on these cut ends. Drench the soil from where the dead roots have been dug out with 2: 2: 250 Bordeaux mixture. Do not allow irrigation water to come into contact with the stem. Avoid deep houng and interculture to obviate injuries to the roots, through which fungus attacks.

MANGOSTEEN

Though one of the most delicious among the tropical fruits, mangosteen (*Garcinia mangostana L.*) is the most difficult fruit-tree to grow on a commercial scale. At present, its cultivation is restricted to a few small pockets on the lower slopes of the Nilgiris at elevations of 300 to 1,000 m and in the Malabar and Tirunelveli districts. Excessive heat or cold, direct sun for long periods, dry atmosphere and desiccated soils are uncongenial to this fruit.

No varieties are recognized. Propagation is mainly done from seed. It is also done by inarching or side-grafting on mangosteen seedlings or on seedlings of Garcinia tinctoria, G. speciosa, etc., but the grafts are difficult to establish in the field. Timely irrigation is necessary to keep the soils moist. The removal of weeds and the replenishment of soil fertility through manuring are very necessary. Definite information on the manurial requirements of the fruit is lacking. As fruits are produced on the terminal portions of the primary and secondary laterals of the past season's growth, pruning is inadvisable. The mangosteen matures (main crop) from august to October and another lighter crop from April to June.

LITCHI

Litchi (*Litchi chinensis Sonn*) is grown extensively in northern Bihar, in the submontane districts of western Uttar

Pradesh and in Punjab. Hot dry winds in summer, when the fruit is ripening, are very undesirable. It grows well in sandy and clayey loams containing large quantities of lime.

Varieties

In Bihar, the recommended varieties include the 'China', 'Purbi', 'Deshi', 'Bedana' and 'Dehra Rose'; in Uttar Pradesh, 'Rose-scented', 'Early Large Red', 'Kalkatia', 'Gulabi' and 'Late Seedless', and in West Bengal, 'China' and 'Muzaffarpur'.

Propagation and Planting

Air-layering is the commonest method. Inarching, budding and grafting are also possible. One-year-old layers are planted nine metres apart in pits during the rainy season.

Pruning

As the fruit is borne on the growth of the previous year, the common practice of breaking off a metre or so of a branch along with the bunch of fruits is all the pruning that is necessary.

Harvesting

Fruit bunches are broken from the tree by hand. A mature tree produces about 110 kg of fruit each year.

REFERENCES

Bose, T.K. Mitra S.K. and Sanyal, D., *Fruits : Tropical and Subtropical,* Third revised edition. Calcutta, Naya Udyog, 2001.

Edwards. A., *Encyclopaedia of Horticulture,* New Delhi, Anmol, 2006.

National Institute of Industrial Research, *Cultivation of Fruits Vegetables and Floriculture,* New Delhi, 2004.

Walter P. Wright, *Encyclopaedia of Horticulture and Gardening,* Delhi, Biotech, 2002.

Yadav, P.K., *Fruit Production Technology,* Lucknow, International Book, 2007

3

Production of Fruits of Arid and Semi-arid Areas

DATE-PALM

Date-palm (*Phoenix dactylifera L.*).Some of the imported varieties of date-palm have been cultivated successfully in Punjab. This tree thrives in light as well as in heavy soils, provided the depth is adequate (2.5 to 3 metres) and the drainage is good. The plant is not affected by frost and requires intense heat in summer for the development and ripening of the fruits. The maxim 'head in fire and feet in water' indicates the conditions that are ideal for date-cultivation. If the water-table is within 3 to 3.5 metres from the surface, mature trees need no surface irrigation. A wet season during flowering and fruit-ripening is limiting factor in the successful cultivation of the date-palm. South-western Punjab, northern Rajasthan and Kutch offer suitable conditions for date growing.

Climate

The date palm is a tree flourishes well under varied sets of climate. However, for proper maturity of fruit, the date requires prolonged summer heat without rain or high humidity during the ripening period. A light shower with prolonged periods of cloudy weather and high humidity may cause more damage than a heavy rain followed by clear weather and drying winds. The mean temperature between the period of flowering and ripening of the fruit should be above 21° C rising to 27° C or higher for at least one month. For successful fruit maturation, nearly 3000 heat units are required. The heat units available in most of the north-western districts of India, counted from the time of flowering, i. e. end February to July, indicate that these are suitable for date palm cultivation. Taking this into consideration, parts of Rajasthan extending from Sri Ganganagar to Anupgarh, Jaisalmer and Barmer and Kutch district of Gujarat will be suitable for date cultivation. The climate of Abohar (Panjab) should be considered marginal because of rainfall (30 cm) and high humidity at the time of ripening (July-August) being the limiting factor.

Soil

The date palm can be grown in soils containing more alkali or salts. It can tolerate such condition better than many other fruit plants. In view of the large investment required to bring a date garden into bearing and maintain it on profitable production, sandy loam soil, 2 or 3 metres deep with good water holding capacity and drainage is most desirable. Date palm can tolerate high soil salinity. In the soil having 4 per cent salt concentrataion, dates can survive well provided the root system does not come in contact with a stratum of soil where the sodicity is more than 1 per cent.

Choice of Varieties

Hillawi : It is soft date from Iraq. This variety is relished both in doka and dang stages. Total soluble solids range between 28 to 42% and astringency in the fruit at doka stage is low or almost absent. It is an early variety yielding good crops and is somewhat more tolerant to rains. Dry dates (Chhuhara) of good quality and cured soft dates (Khajoor) can be prepared. Its fruit at dang stage (fresh form) is very delicious. Average yield ranges from 50 to 80 kg per palm.

Khadrawi : It also originated from Iraq. This variety proved successful under Abohar conditions. The trees of this variety are comparatively less tall and and yield good crops. It can be used both for dry dates and soft dates. The yield ranges from 40 to 70 kg per palm.

Shamran : It is a mid-season cultiver tolerant to high humidity. The variety yielding prolific crops. The fruits can be cured into good quality. Chhuhara and also softened with salt treatment. Fruit is medium to large, oblong to oblong-oval and yellow at doka stage. Its yield per palm is 40 to 70 kg.

Medjool : The variety originated in Morocco and has large size fruits. It is late in ripening and has proved particularlt good in preparing dry dates of attractive bold size and good quality. Its yiled is 35 to 50 kg per palm.

Barhee : It is soft date from Iraq. The variety has proved extremly good for table use in fresh form (doka stage). The doka fruit contains about 32% TSS, has golden yellow colour and has a very pleasant taste because of low astringency and high pulp content. The trees of this variety are prolific in yields and the fruit is ready for eating late in the season i.e. first fortnight of August at Abohar. Its yield ranges from 60 to 110 kg per palm.

Hayany : It is soft date originated from the UAR. The berries have attractive deep colour. The fruits could be consumed as fresh in the doka stage as ripening does not proceed further under Abohar conditions. The yield per palm ranges from 30 - 40 kg.

Zaidi : It is a mid season variety, slightly tolerant to rain or high humidity. The fruit is small to medium, obovate and yellow at doka stage.

Propagation Techniques

The date palm are always propagated throughsuckers (offshoots) for commercial plantation. The suckers usually arise near the ground around the trunk. The offshoots arising at a higher level from the ground on tree trunk and with no roots give very poor survival. These are separated from the mother palms during March or August-September. The leaves are trimmed off 4 - 5 days before their detachment. The inner leaves should be cut back by one-half and the outer ones by two-thirds. The tender young unopened leaves near the central bud and parts of the bare stalks of the old leaves necessary to protect the bud, should be kept on the offshoots. The outer whirl of leaves may be fastened to afford protection from heat and cold to the central bud also facilitate detachment and transport.

The suckers weighing approximately 25 to 30 kg and well rooted should be removed. Expose the point of detachment and set a chisel (12 cm wide and 19 cm long) sharpened at the face and welded to a 150 cm long handle of 4 cm diameter. Drive in the chisel by hammering at the other end.

Planting Operation

The offshoots are planted 6 to 7 metres apart in the already prepared pits of 1 × 1 × 1 m size immediately after detachment. In this way 202 to 275 plants/hectare will be accomodated. If transported from long distance, the offshoots may be kept in the straight. The soil around the offshoots should be pressed firmly. The field is irrigated immediately after planting. Thereafter, frequent light irrigations are given to keep the soil always moist. At Abohar, the offshoots removed in March survived better in direct field planting than those removed in August-September.

Orchard Cultural Practices

Irrigation

Date palm is known as drought resistant fruit tree and is

able to survive for long periods without irrigation. However, continuous drought condition retards the growth of the plant. If water is available, date palm uses it lavishly. Date palm is highly tolerant to excessive irrigation and floods. Continuous stagnation of water or waterlogged conditions are injurious for its growth. To maintain maximum growth, the root zone up to 2 - 3 m should be kept moist and not allowed to dry. Light but frequent irrigation should be given after planting.

The full grown trees on the lighter soils are usually irrigated 7 to 10 days during mid-summer and every 15 to 30 days during winter. Such soils requires 2.75 to 3.75 acre-meters of water per year and 30 to 45 acre-centimeter per month is needed during summer. On heavy soils, half the amount may be enough. Each irrigation of 10-15 acre-centimeters is usually sufficient. Irrigation is withheld when ripening starts to facilitate harvesting, hasten fruit ripening and to reduce fruit drop caused by high humidity.

Manuring and Fertilization

Generally the farmers do not apply manure to the date palm. The trees certainly respond to manuring as indicated by increased vigour and growth. The results of studies conducted at Abohar shows that 25 to 50 kg of farmyard manure and 1 to 2 kg ammonium sulphate per palm should be added. The quantities may be reduced if some leguminous crop has been grown in the date palm. Farmyard manure is applied during December-January. The nitrogen dose should be given a fortnight before flowering i.e.in the first week of February.

Intercropping

Intercropping in date palm with suitable crops bring good income and also improves the fertility of the soil. During the first few years, intercropping can be practised with no shortage of irrigation. Intercrops such as gram, peas, mash, moong, moth, senji and lentil can be sown during summer. Intercropping of some vegetables in plantation located near the cities, can be practised if sufficient irrigation and manuring facilities are available. The filler trees like grapes can be tried with good success.

Fruit Thinning

Excess load of fruit may cause shrivelling of berries, breaking of spathe stalks, more damage due to rain and humidity, delay in ripening and alternate bearing. It also reduced size and produce poor quality of fruit. It is, therefore necessary to keep only optimum quantity of fruit and thin out the rest. This is usually accomplished either by reducing the number of fruits on each bunch and or by removing some of the bunches. The number of fruit that a palm can safely carry depends on the cultivar, age, size and vigour of the palm and number of green leaves on it. Under normal conditions, 1 - 2 bunches in the 4th year and 3 - 4 bunches in the 5th year may be left. Normally 8 - 10 bunches per palm are retained in India. Small, defective and broken bunches should be removed.

In short-stranded varieties like Khadrawi, the strands are generally cut back to even up the bunch from the top. Most of the fruit thinning is done by the removal of half to two-third of strands from the centre. In the long stranded varieties like Deglet Noor, one-third to half strands are cut in similar way as in Khadrawi, in addition, strands are also cut back to remove about one-third of the flowers.

The desirable number of fruits to be left is between 1300 and 1600 per palm depending on the variety. The per cent thinning is generally done 40 - 50 in Khadrawi, 50 - 55 in Hallawi, 50 - 60 in Zaidi and Barhee. Ethephon 100 - 400 ppm after 10 to 30 days from fruit set was found effective in fruit thinning of cv. Hayany. The biennial bearing habit of the treated palms were found to reduce by ethephon treatment. It also advances the ripening of fruit.

Harvesting and Processing

Harvesting of dates generally depends upon the weather conditions. The rain and high humidity in the atmosphere at the ripening time is a limiting factor in its cultivation. When the climate is favourable, it is preferable to leave the fruit of most varieties on the palm until it reaches the stage of maturity at which it is to be consumed or stored. The changes associated with ripening and the period during which the fruit may be consumed extend from the peak of the khalal stage, when the fruit has its most intense red or yellow colour and maximum weight, to the final tamar stage, when it has lost the greater part of its moisture content and will keep without special attention to storage. Most of the people like to eat the fruit in khalal stage. At least two varieties of dates Hillawi and Barhee, are liked most for eating in khalal stage. Dates are hand-picked at the stage of maturity. All the dates in the same bunch do not ripen at same time, it has been the practice to make several pickings to harvest the fruit during a season. Sometimes, when the season is favourable and more than 70 per cent fruit is ripe, the entire bunch is harvested.

The research conducted on curing of dates in Panjab (Abohar) have revealed that the rain and high humidity in the atmosphere at the time of ripening (July and August) do not allow the ripening process on the tree to proceed satisfactorily beyond doka stage. If the fruit is retained on the tree, there is checking, splitting and rotting. High humidity and rainfall also cause the fruit to drop. Therefore, the crop has to be harvested at doka stage. Though in comparatively drier seasons partial crop on some trees may become dang (mellow and soft) but the quantity of such fruit is limited. The berries are removed and graded manually on the basis of size and colour.

Scientists working at Abohar (Panjab) have developed a technique to transform satisfactorily the date fruit at doka stage into dry dates (Chhuhara) of good quality. At least four varieties Hillawi, Khadrawi, Shamran and Medjool have yielded very good product. Chhuhara obtained from Medjool, which is a large-sized variety, compare very well with the high quality Chhuhara imported from Middle East countries. The technique developed involves immersion of frut at doka stage in boiling water for 6 to 8 minutes and then drying either in temperature-controlled oven (air-circulation type) for 80 to 120 hours at 48° to 50°Cor in the sun for 10-15 days if weather is dry. Thus, an average of 45 per cent fruit product is obtained. Fruits at advanced doka stage or when they attain one-fourth, one-half or full dang (the berries become mellow and soft starting from the distal end) can be converted into soft dates (khazoor) of good quality by drying either in the oven at 40°C or in the sun. Thus, for soft dates, only drying the berries at partial or full dang stage is required with no other treatment and this way a final product of soft Khazoor ranging from 50-60 per cent is obtained.

Hillawi and Barhee yielded 50-100 kg fruit per palm at doka stage. The average yield from Khadrawi and Shamran is obtained 40-70 kg fruit per palm. The yield from Medjool palm is upto 50 kg but its fruits are very large in size.

Artificial ripening of fruit at doka stage to transform it into dang stage was also attempted at Abohar. For these studies, berries were treated with 0.5 per cent to 2.0 per cent common salt (sodium chloride) and similar concentrations of acetic acid in combination with 1.0 per cent salt. The fruits to be treated with salt were spread on polythene sheet and requisite quantity of salt was applied by rubbing and smearing uniformly on the berries. The fruits which were given acetic acid plus salt treatments, were first dipped in solutions of desired acetic acid concentrations for 2 minutes followed salt application by the method described above. Each treated lot was packed into wooden boxes lined with old newpapers and packed in laboratory at room temperature. The boxes were opened after 24 hours. From after 24 hours. From these studies, it was found that 2 per cent salt, 60 to 70 per cent of the doka fruit were transformed into dang. However, the dang obtained by this treatment was not as good in taste as that of naturally ripened on the tree, but still it was edible and generally acceptable on account of its having lost the astringency. But such products cannot be stored more than 24 hours and, as such, should be consumed to as early as possible.

Insect-pests

The most important pests of dates are follows:

Date Stone bettle

It attacks unripe fruits and the adult penetrates the fruit and reaches the stone by constructing a direct circular hole through the pulp and both adults and grubs feed inisde on the pulp and results in the drop of unripe fruit. Hillawi and Khadrawi, which ripen early, are more suscepitble to this pest. Spraying the plants with 0.075% per cent Malathion has proved effective for its control.

Nitiduled bettle

The larve penetrate the fruit and eat the inner portion of the pulp. Their attack is followed by fungal decay. The adult is brownish yellow, oval, slightly oblong, 2.3 + or - 0.13 mm long and 1.16 + or - o.21 mm broad. Initially, the dropped fruits are attacked. Then the peat spread to fruit bunches on the tree and cause the fruit to drop. The pest over winters as pupae in the soil. Spraying wiht Malathion (0.075 per cent) gave satisfactory results. Regular removal of dropped fruits and their destruction also helps in minimising the effect.

Cigar hoeing bettle

It is reported as pests of stored dates in India. For the

control of these pests, fumigate the dates and keeping material with Methyl bromide at a dose of 1/2 kg/27 cubic meter giving an exposure of 24 hours.

Termites

They feed on the roots of the newly transplanted suckers which may ultimately cause them to wither and die. As a preventive measure, 10 per cent BHC mixed in each pit @ 30 g before planting is recommended.

Diseases

Much less work has been done for the investigation of diseases of date palm in India. Studies conducted at Regional Fruit Research Station, Abohar, have found two diseases.

Graphiola leaf spot

It is also called as false smut or palm leaf pustule. It is a serious disease. This disease is more acute in the areas where humidity is more. Numerous hard black smut like pustules are seen scattered developing epidermily on the upper and lower surfaces of leaflets and also on rachis. The pustules at maturity consists of round, hard, black cups about 0.50 × 0.25 mm. These cups contain yellow spores. Older leaves are more affected, they loose their chlorophyll and dry up. For its control, remove and burn the affected leaves to check further contamination and spray the trees with 2 : 2 : 250 Bordeaux mixture (2 kg copper sulphate + 2 kg unslaked lime mixed in 250 litres of water.)

Fruit rot

It often causes considerable loss when humid weather occurs during the ripening season. Under such conditions, various fungus may develop in the fruit and cause spotting, dropping and rotting. The damage may be reduced by better ventilations of the branches and protecting the fruit from rain. In addition, spray the bunches with a mixture of 5 per cent fahana (ferlic dimethyl dithiocarbamate) in sulpher.

PHALSA

Phalsa (*Grewia asiatica L.*) Phalsais a small bush which grows all over the country except at higher elevations. In Punjab, Haryana, Uttar Pradesh and Andhra Pradesh, it is grown commercially. The fruit is berry-like, globular, smooth, deep reddish brown and comes in the market in May and June.

Climate

Phalsa relish distinct winter and summer for best growth, yield and quality. In regions having no winter, the plant does not shed leaves and produce flower more than once, thus yielded poor quality fruits. Full grown plants can tolerate freezing temperature for a short period. The plants can tolerate temperature as high as 44°C . High temperature during fruit development favours ripening of fruits. At flowering time, clear weather is needed, whereas rains at that time affect fruit-setting adversely.

The phalsa is one of the hardiest fruit plants and can be grown successfully throughtout Punjab and Haryana. Being deciduous in habit, it can withstand frost quite well. However, care should be taken to select sites which are not subject to very low temperatures during winter. It can also grow successfully in hot and dry climatic conditions, such as those prevailing in the districts of Ferozepur, Bhatinda and Sangrur of Punjab. It can withstand occassional drought and water shortages better than fruit plants. However, for raising profitable crop of phalsa, satisfactory arrangements for irrigation need to be made.

Soil

The phalsa is not at all fastidious in its soil requirements and can grow quite successfully in almost every kind of soil. Around Amritsar, where the soils are comparatively heavy, its plants grow well and bear bumper crops. At many other places such as Ludhiana, it yields well under lighter soil conditions. Though the ideal soil for growing phalsa is the rich loamy type. Under water logged conditions plants become chlorotic and make poor growth.

Proper soil drainage is another factor which should be kept in mind. However, soils where water stagnates for many days during the rainy season or those which have a poor sub-surface drainage and are water-logged should not be selected for commercial cultivation of phalsa .

Choice of Varieties

In phalsa, no distinct cultivar is identified. Some growers have, however, given names as Local and Sharbati. Two distinct types i.e. Tall and Dwarf were recognised at Hissar. Dwarf type was found to be more productive.

1. *Tall type* : Plant height is 4.5 m, leaf size 20 ´ 18 cm. Leaves are light green in colour. Average fruit size is 2.07 cm and weighing 0.48 g. Average yield per bush is 5.2 kg. Pulp is 81.5 per cent, juice 5.4 per cent. TSS 14 per cent and Acidity 4.64 per cent.
2. *Dwarf type* : Plant height is 3.4 m, leaf size 18 ´ 15 cm. Leaves greenish white in colour. Fruit size is 2.26 cm, weight 0.54 g. Average yield is 3.5 kg per bush. Pulp 60.3 per cent, juice recovery is 34.6 per cent. TSS 12.1 per cent with acidity 3.63 per cent.

Propagation

For raising the phalsa seedlings, large sized, purple black coloured fruits are collected when the crop is ready in the end of May or beginning of June. After extracting from the fruits, the seeds should be washed and dried under shade. These seeds are sown in raised beds when there is sufficient humidity in the atmosphere. Rainy season (July-August) is the best time for sowing of seed. The seeds on raised beds are sown in lines which are 10 - 15 cm apart from each other. In the rows, the seed should be sown 4 - 5 cm apart and 1.5 to 2 cm deep. The seeds should be covered with sand or light soil mixed with well rotten and dry farmyard manure. The beds should be watered regularly with water, but too much watering should be avoided as the excessive moisture conditions cause poor development of the plants. The beds should be kept free from weeds. Initially, one hand weeding is recommended, otherwise the roots of small plants get damaged if weeding is done with khurpa etc.

When the seedlings come out and have made 5 - 7 cm growth, light dressing of calcium ammonium nitrate or Ammonium sulphate at the rate of 50 g/sq metre of nursery area is applied. This encourage rapid growth of the seedlings. Seedlings are ready for transplanting in the following winter i.e. during January-February.

Propagation through cuttings : The phalsa plants can also be multiplied through hard-wood cuttings. The cuttings should be prepared during December-January and kept for callusing. The use of root promoting harmones such as Indole butyric acid @ 100 ppm enhances the success of rooted cuttings.

Planting Operation

Land is prepared well before the plants are set in the field. When there is enough time, a cover crop such as guara, jantar or senji may be sown in the soil. After these crops produce enough vegetative growth, they should be burried into the soil. Before planting, the orchard site should be properly laid out according to the square or the hexagonal system. Half metre deep pits or half metre diameter should be dug and refilled with a mixture of top-soil and well rotten farmyard manure in the ratio of 1 : 1. To save the plants from possible attack of white ants add 30 g of BHC 10 per cent dust to each pit. After this, the field is thoroughly irrigated and transplanting is done when the soil is in friable condition.

The phalsa plants are spaced at various distances in different regions of India. Eight to twelve months old seedlings are better for planting in the field. The plants are spaced at 1.5 metres apart by which 4400 plants are accommodated in one hectare by the square system of planting.

The phalsa plants should preferably be transplanted in the field during January-February before they start new growth. The plants being dormant at that time, they can be lifted from nursery with bare roots. However, for transplanting during August-September, the seedlings have to be lifted from the nursery alongwith earth balls. Transplanting in the rainy season, is, therefore, a little more cumbersome and riskier than that done in the spring, when the plants are dormant.

Flowering and Fruiting

Flowering in phalsa starts from February-March and continues till May. The first flower to open is at the base. Flowers are borne in the axil of leaves. The flower are mostly cross pollinated and honey-bee seems to play major role in pollination.

The flower buds become plumpy before anthesis. The first sign of anthesis is the appearance of a slit in sepals at the base of the bud. The slit widens and at first only one sepal falls apart. The other sepals fall one by one and the whole process of flower opening is complete within half an hour. The dehiscence of anthesis in phalsa takes place before the flowers are completely open.

Orchard Cultural Practices

Irrigation

Phalsa is regarded a drought resistant fruit plant. However, it can be observed that in Punjab, Haryana and other neighbouring Northern Indian regions, irrigation is essential for securing high yield of better quality fruits. Its plants do not need any irrigations till January. Adequate supply of irrigation water at regular intervals specially during flowering and fruiting periods is very essential for ensuring better health of plants and more profitable yields. A light irrigation should be given after pruning and fertilization of the plants. During the early part of the spring, when the new growth starts and the temperature is usually low, the irrigation interval may be kept at about 2 - 3 weeks. The demand for water increases a great deal after the middle of April and consequently the irrigation interval may be reduced to 7 - 10 days. At this time, the plants should not be allowed to suffer from water shortage; as it would adversely affect the yield and quality of fruit. Regular irrigation should be applied to the plants till the fruit is harvested and the monsoon sets in.

As the roots of the plants are active during winter, adequate soil moisture should be maintained in the root zone. During winter, one or two light irrigations may suffice if there are no rains.

Interplanting

Many fruit plants grown in the plains of North India such as mango, litchi and pear comes into bearing on commercial scale very late. These fruit plants have a slow rate of growth and the space between the tree rows remains unoccupied for several years. The growers can, therefore, utilize the empty space between the permanent fruit trees for raising some quick-growing fruit plants like phalsa. Phalsa has another advantage over other quick-growing fruits suitable for interplanting that it is kept low-headed by severe pruning every year and does not cause overcrowding in the orchard. Phalsa being a deciduous plant, does not require much water during the dormant period (winter months). Phalsa should be interplanted only in the middle of the permanent tree rows. As soon as the permanent trees come into bearing, the phalsa plants should be uprooted, so that the main fruit crop does not suffer.

Manuring and Fertilization

Generally, phalsa is planted on comparatively poor soils. Some cover crop such as gurar, jantar or senji should be grown in the field where phalsa is to be planted. This practice adds sufficient quantity of organic matter to the soil and enriches it.

In order to get profitable crops of good quality, full grown phalsa plants should be given 10 - 15 kg of well rotten farmyard manure, soon after planting. Nitrogenous fertilizers (about 1 kg of CAN or Ammonium sulphate per bush) should also be applied, preferably in two split dose-one at the time of flowering and second after fruit-setting. Higher yield of phalsa can be obtained by application of 100 kg N, 40 kg P and 25 kg K per hectare, respectively. Trials at Rajsthan Agricultural University, Udaipur revealed that application of NPK @ 100, 40, 25 kg per hectare gave higher yield.

Zinc and iron were found to influence berry size and juiciness in phalsa. $ZnSO_4$ @ 0.4 per cent at prebloom stage and after berry set improved the juice content. Ferrous sulphate at 0.4 per cent alone or in combination with zinc improved the berry size.

Pruning

The general practice of cutting back the phalsa plants to the ground level every year during the dormant period is not in order. Investigations carried out on this aspect have shown that the phalsa plnats pruned to a height of one metre during January-February produce a greater number of new shoots than those which are pruned to the ground level. The growers are, therefore, advised to prune their phalsa plants at a height of one metre from the ground level.

The phalsa plants are rather slow in shedding their leaves winter. The best time for their pruning is when the plants have shed their leaves and in all cases the operation should finished well before the start of new growth.

Use of Growth Regulators

Growth regulators have proved effective in increasing the fruit set and yield in phalsa. GA3 at 10 ppm increased the setting of fruit and yield appreciably. CCC at 250 ppm sprayed twice at an intervalof 7 days after 50 per cent fruit set increased the fruit size. Ethephon at 1000 ppm resulted in maximum ripening in 5 days after application. SADH 1000 ppm at early stage of plant growth reduced the plant height and brought improvement in yield and fruit quality. Spraying of GA3 at 60 ppm once at the beginning of flowering, another after 15 days and ethrel at 1000 ppm when ripening of berries had just started, increased the fruit retention percentage and yield. They also reduced the harvesting span and increased the TSS of the fruit.

Harvesting of Fruit

The phalsa plants begins to bear fruits in the second year. A good commercial crop is usually obtained during third year. In the Punjab and Haryana, the harvesting season of phalsa fruit starts by the end of May and lasts till the end of June. The fruits should be picked when the colour has changed to a deep reddish brown and the pulp tastes sweet. Several pickings are necessary as all the fruits do not ripen at one time. The fruit-picking is usually done on alternate days.

Insect Pests

Psylla: A tiny insect causes severe damage to the phalsa foliage and young shoots. Its infestation over a long period of time results in the deformation of floral and vegetative parts. The plant growth is greatly arrested and gives a sickly appearance. Fruit set is severally reduced resulting in great financial loss to the growers.

The pest can be controlled by spraying one kg of carbaryl (Sevin 50% WP) in 500 litres of water or 300 ml Malathion 50 EC or 300 ml of Rogor 30 EC in 300 litre of water per acre as soon as new growth starts in March.

Diseases

Brown spot

The disease is caused by fungus Cercospora grewiae. The brown spot of phalsa is quite wide spread in the Punjab and is very severe from June to August. It results in premature leaf-fall when the attack is severe. The disease first appears as tiny lesions on the upper and lower surface of the leaf. In the beginning, these lesions are covered with a white mass of fungus. Gradually, the lesions enlarge and become reddish brown to dark brown. Many times, several lesions coalesce to form big spots which are very conspicuous and may cover a large part of the leaf.

To control this disease, the leaves and prunings should be collected soon after removal and destroyed to check the fresh infection of the new leaves, the plants should be sprayed with Bordeaux mixture 2 : 2 : 250.

Pinspot of phalsa

The disease is caused by phyllosticta grewiae. The disease can appear any time during the growing season of phalsa and may cause considerable damage to the foliage. Small brown to dark brown, circular to irregular pinspot like lesions appear on the leaves. This disease can be controlled by sprays given above for the brown spot of phalsa.

Irrigation

The banana-plants require very heavy irrigation. Irrigation is given in most places once in seven to ten days. Stagnation of water in the soil is not very congenial to the proper growth of banana and, hence, the drainage of soil is also essential.

Phalsa is commercially raised through seeds. The seeds are sown in well-prepared nursery-beds, 5 to 7 cm apart, in rows normally kept at 25 to 30 cm apart. The best time of sowing is July-August. The seedlings, thus raised, are ready for the transplanting in the following January-March or July-August.

A Phalsa plantation is raised by transplanting the seedlings, 3-4 metres apart, according to the square system of planting. February is considered the best month for this purpose.

Phalsa is not ordinarily manured. However, with about 10-15 kg of farmyard manure per bush, better returns can be obtained. It can stand drought and does not need frequent watering.

Pruning is an essential operation in phalsa cultivation and is done in January. Severe pruning is practiced by some growers almost to the ground level.

Harvesting is done during May-June and since the fruits ripen gradually, several hand-pickings are necessary. The average yield per plant is 5-6 kg.

CUSTARD-APPLE

Custard-apple (*Annona squamosa L.*) or sitaphal is a small tree seldom more than 4.5 metres high and is both cultivation and found growing in a semi-wild state throughout Peninsular India. A dry-and-hot climate suits it most. It flourishes in lighter soils, and grows well even on the slopes of hills, but cannot stand cold or frost.

This fruit-tree is propagated through seeds and by grafting. There is considerable variation in the fruit-trees when propagated from seeds. 'Bullock's-heart (ramphal provides a suitable rootstock for grafting by inarching or budding. Grafted plants give fruits earlier (in about two years) than seedlings which take three to four years. The plant flowers from April and November. The yield per tree varies from 25 to 30 kg. Fruits are mostly consumed locally, but they can be transported over long distances if they are picked before they are fully ripe and are packed in a single layer in well-ventilated wooden boxes with soft padding material.

BULLOCK'S-HEART

The Bullock's-heart (*Annona reticulata L.*). is an allied fruit which is rarely cultivated. It prefers a heavier soil than custard-Apple and is less resistant to cold and frost. It is easy to propagate by inarching. It gives fruits from January to May and the yield goes up to 45 kg per tree.

Cherimoyer is the best of the family and performs best in subtropical climate. It is propagated by inarching on Bullock's heart rootstock. The tree bears fruits from the sixth year onwards and gives about 100 fruits per year. Attemoya is a cross between custard-apple and cherimoyer and is similar to the latter in its growth requirement.

AONLA

Aonla (*Phyllanthus emblica L.*) is indigenous to India and the tree is characterized by very small leaves which are attached to branchlets in such a way that each branchlet looks like a compound leaf. The tree attains good height. It is quite hardy and can be grown without much care in all types of soil, except very sandy. The 'Banarsi' aonla is the most important cultivar which is propagated vegetatively. In the District of Pratapgarh of Uttar Pradesh, there are many commercial orchards. The important varieties grown are 'Banarsi', 'Chakiya' and 'Pink-tinged'. These are vegetatively propagated and highly valued unlike the seedling-trees which produce unmarketable fruits.

The commercial practice of propagating aonla is by shield-budding which is done in June. The two-year-old seedling aonla is used as rootstock. Budded aonla plants are planted in the fields during rains at a distance of 11 metres each way. It can also be planted as a windbreak around an orchard. The budded plants will commence bearing from the 10th year. The vegetative growth of the tree continues from April to July. Along with the new growth in the spring, flowering also commences. The flowers are of two types : i male, and ii female. Fruits are fully mature by January-February. Under conditions in southern India, the fruits are found throughout the year at one place or another.

FIG

Fig (*Ficus carica L.*) has never been grown extensively in India. Maharashtra formerly had more than 500 hectares, but largely because of diseases, this figure has decreased to less than 125 hectares. Southern India reports somewhat less than this area, and there are a few fig-groves in Hyderabad and in other parts of the country. The types are not of the best quality and it is likely that they are hybrids between the European and other species. The climate of most parts of India seems to be suitable for the cultivation of fig. Rain at the time of fruit ripening is undesirable.

Varieties

There is as yet little basis for recommending specific varieties for different parts of India. The variety most commonly grown in Maharashtra and southern India is known as the 'Poona' fig.

Propagation and Planting

The plants are almost entirely grown from cuttings. Budding and grafting are easily done. Several species of Ficus are compatible as rootstocks. Cuttings made in winter are ready for planting out in the rainy season. Some prefer to plant the cuttings in situ. The spacing of fig-trees depends on the size they are to reach and this depends mainly on the type of pruning. Generally, a spacing of 4.5 to 5 metres is adopted.

Culture

Tillage to keep down weeds is necessary and irrigation during the dry season when the fruits are growing and maturing is desirable. The tree is deep-rooted and drought-resistant, but does not bear well under conditions of water deficiency. A good supply of nitrogen is also required for heavy bearing, and the use of farmyard manure is also recommended. A fertilizer mixture containing 0.225 kg of N, 0.45 kg of P and 0.45 kg of K per tree may be applied.

Pruning

The fig-tree can stand very heavy pruning. Most of the fruit is borne on the growth of the current year. Some annual pruning seems desirable, but the most efficient type in each of the areas where the fig is grown in this country is not known.

Harvesting

The ripe fruit is delicate and must be harvested carefully and used within a few days. Where figs are grown for drying, they may be allowed to fall from the tree on clean dry ground.

JUJUBE

Jujube, ber or bor, is estimated to occupy about 4,000

hectares mainly in the northern, eastern and western parts of India. Almost any soil is suitable, even if it contains enough alkali to prevent the growth of some crops. The tree can stand prolonged drought and also water-logging.

Climate

The ber is a hardy fruit tree and can grow successfully even under unfavourable climatic conditions, where most other fruit trees fail to grow. It can be grown up to a height of 1,000 meters above sea-level, beyond this it does not bear well. Ber relishes hot and dry climate for its successful cultivation, but the trees need adequate watering during the fruiting season. Excessive atmospheric humidity is considered a limiting factor for satisfactory fruiting. Frost during winter does not have much effect on the tree. It can very well withstand hot and dry weather during summer months of May - June, its tree enters into dormancy by shedding its leaves. New growth starts in July with the advent of rains and the growth period continues till the middle of November when it is inhibited with the onset of cold weather.

Soil

Ber provides a good scope for cultivation on soils which have so far been considered marginal or even unsuitable for growing other fruits. Its tree developes a deep tap-root system within a short period of growth and as such adopts itself to a wide variety of soils. Ber is well known for its ability to thrive under adverse condirions of salinity, drought and water-loggin. The research work conducted at the Punjab Agicultural Universityhas revealed that ber can flourish even in soils with pH as high as 9.2. Ber cultivation can succeed even under constraints of irrigation and fertilization, however, for good tree growth and yield, deep sandy-loam soils with neutral or slightly alkaline reaction and good drainage are more desirable.

Promising Varieties

Umran: This vsriety is cultivaed in a large scale in Punjab and Haryana. The fruit is large, oval in shape with a roundish apex and has an attractive golden yellow colour which turns into chocolate brown at fully maturity. The fruit is sweet, with 14-19 per cent TSS and has pleasant flavour and excellent dessert quality. It is a prolific cropping variety, yielding 150-200 kg of fruit per tree. The fruits ripen late from second fortnight of March to mid-April and have a good keeping quality.

Sanaur-2 : This is a selection from Sanaur - a small town near Patiala, which is known for ber cultivation. The fruit is large and oblong with a roundish apex. On ripening, fruits attain a light yellow colour and TSS of 18-19 per cent. Like Umran, it is also a prolofic bearer-yielding about 150 kg fruit per tree. It is a mid-season variety, ripening during second fortnight of March under Panjab conditions and has been found fairly resistant to powdery mildew disease.

Kaithli : This variety is a selection from Kaithal in Haryana. The fruit is medium in size, oval in shape and has a tapering apex. Fruit pulp is soft and sweet with TSS of 14-16 per cent. Fruits ripen in the second fortnight of March to first week of April. The average yield is 120 kg fruit per tree. This is an excellent variety but appears to be more susceptible to powdery mildew disease.

Z.G.-2 : The fruit is medium in size and roundish in shape with smooth skin. The fruit pulp is soft with an excellent sugar-acid blend. When ripe the fruits attain light yellow colour and TSS of 15-16 per cent. The average yield amounts to 150 kg fruit per tree and the ripening time extends from second fortnight of March to first week of April. This variety is recommended for growing for local markets only. It is less susceptible to powdery mildew.

Sanaur-5, Gola, Sanaur-4, Chhuhara and Laddu are other important varieties of ber, which have good taste and bears heavily.

Planting and After Care

Planting : The ber plants can be planted in February-March and again in Augast-September, but the latter season of planting gives a better success. Recently, Punjab Agricultural University has also recommended the bare-rooted planting of ber during December-January.

One metre deep pits of one metre diameter should be dug and left exposed for one month before actual planting. Pits should be refilled with a mixture of top-soil, about 20 kg well-rotten farmyard manure and 1 kg superphosphate per pit. To avoid attack of white ants 30 g of BHC 10 per cent dust is added to each pit. The refilled pits should be about 2-3 cm higher than the ground level and irrigated thoroughly, so that the loose soil settles down firmly. A plant may be set in the centre of a pit with the help of planting board, maintaining the same level of soil at which it stood in the nursery. The budding point should remain about 15-20 cm high from the ground level. The plants should be immediately watered after plantation. They should be irrigated after every 4-6 days atleast during the first two months and subsequently after every 7-10 days for another 3-4 months or until the plants get fully established.

The grafted ber is spreading in habit and grows into a big tree. The tree requries proper spacing for its healthy growth and fruiting. For obtaining good income, ber palnts should be planted 7.5 metres apart in square system (from row to row and plant to plant) thus accommodating 180 trees per ha.

Lifting plants from the nursery : The budded ber plants should be transplanted during February-March or Augast-September. The plants are dug out from the nursery with good-sized earth-balls so that their root-system is not much distributed. The plants should be packed carefully to keep the earth-ball intact. The earth-ball should be kept moist throughout the period of transportation and upto plantationso that the roots of tree do not dry up in the intervening period.

While lifting bare-rooted plants, the leaves and shoots of the budded ber plants are removed before uprooting them from the nursery. These plants are also headed back at a height of 60-75 cm from ground level. Then, these plants are dug from the soil with the help of of Spade and Khurpa to keep maximum feeding roots with them. These plants are tied loosely in bundles and are wrapped in moist rice straw. Such plants can easily be trained according to modified central leader system. The bare-rooted plant should be lifted from nursery from mid-December to early-February.

Planting as Windbreak : Besides regular ber plantations, the tree can also be successfully planted as a windbreak around the commercial orchards to provide an effective wind screen. The tall-growing tres like Safeda, Arjun and Simbal shouls be interplanted with ber, being a low-headed tree.

Rootstock and Propagation

Seeds of Katha ber (*Zizyphus mauritiana Lamk*) are generally used for raising rootstock, which are easily available from the wild growing trees and possess the qualitiesof a good rootstock. The ber plants should be budded on Zizyphus mauritiana for higher fruit yield. Ber plants raised on semi-vigorous rootstock Zizyphus mauritiana (Coimbatore) can profitably be planted at a closer spacing of 6 x 6 m. Umran trees grafted on this rootstock show a spread of 6 metre as compared to the recommended rootstock viz. Zizyphus mauritiana (*Elongated Dehradun*) on whom the tree spread to an area of 7.5 metre. Thus about 50% more plants/ha can be accommodated with over 20% increase in yield of equally good quality fruit.

Seeds of Mallah ber (*Zizyphus numularia*) can also be used as rootstock. The seedlings of Mallah ber are slow growing and become buddable after longer period than the seedlings of Katha ber.

Raising of Rootstock : The germination of ber seed is quite a difficult process on account of the stony nature of the shell (endocarp) which contains the seed. A large percentage of seed stones are non-viable and require elimination at the time of sowing. Seed stones collected from dropped fruits contain 50-70 percent non-viable seeds. Seeds should be dipped into a salt solution of 17-18 percent concentration for 24 hours before sowing. The flaoting seeds should not be sown as these are generally non-viable. The ber seeds can also be sown by cracking the hard shell (endocarp). They germinate rapidlly in about 8-10 days. However, for commercial purpose, stones should be sown as such to avoid injury to embryos. To get best root-stock material, it is important that seeds should be collected only from healthy and vigorous-growing wild ber trees.

The ber seedlings raised from Katha ber stones, which are sown during March-April, after fresh extraction, in well-prepared nursery field at a distance of 15 cm in rows 30 cm apart. Germination starts in about 3-4 weeks and seedlings make a rapid growth. The seedlings should be trained to the single stem. Nearly one-fourth of the seedlings attain buddable size of a lead pencil by Augast, while the rest are ready for budding by April next.

Budding : The propagation of ber by budding is the most successful method. Both Shield or T-budding and ring-budding methods are employed but the former is preferable, because it is easier to perform. Budding operation should be done when there is proper flow of sap in the stock to be budded. Shield-budding is done during March-April or August-September, but it has been found that August-September budding gives a far better success. The buddlings make growth at a very fast speed. Plants budded in April usually become saleable in August-September, where as those budded in rainy season are ready for planting by February-March next. The ring-budding is preferable during June-July when the new growth starts. Shield-budding done during August-September has given success of 75-81 per cent, whereas budding in April has given a little success. The highest budding success is also achieved in June.

Selection of Stock and Scion: The stock seedlings should be healthy and vigorous and it is allowed to grow as a single stem only. The budding is done when the stock stem has attained the thickness of a lead pencil.

It is essential that bud sticks are taken only from selected trees which are known for bearing a heavy fruit crop of good quality. The mother plant from which the scion bud-sticks are obtained for budding should be healthy, vigorous, free from diseases and insect-pests and should be true-to-type. Two to three months old shoots with plump buds should be selected. Bud-sticks, about 30 cm long, are cut from the selected mother plants. For sending to distant places, about 20 cm long bud-sticks should be taken. The leaves of the bud-sticks should be cut away, keeping the leaf stalks attached to the buds. These should be tied in small bundles and wrapped in moist piece of cloth. The bud-sticks should be kept moist till they are used for budding. The bud sticks can be kept for 2-3 days in good condition.

Flowering and Fruiting

The whole period of bud development is divided into eight stages. It takes 21 days to 22 days for passing through various stages of bud development. The flowering period lasts for about two and a half months from September to November. The peak of flower anthesis is at 6 A. M. in Sanaur-2, Sanaur-5 and Chhuhara ultivers and 2 P.M. in Thornless, Khaithli and Umran cultivars. The dehiscence of anthers starts just after anthesis and completes within 4 hours. Pollen grains are highly viable - the viability ranging between 87 to 91 percent in different cultivers. Pollen grain germinability is also quite high to 36 - 48 per cent. The peak receptivity in stigma is found on the day of anthesis. It does not set any fruit by self-pollination, thereby shows self-incompatibility.

Fruit-setting starts in second week of October and continues upto first fortnight of November. The ber fruit reaches to ripe stage in about 180 days after fruit-setting. The fruit growth in terms of lenght and diameter shows three distinct phases and follows a pattern of 'Double Sigmoid' curve.

Orchard Cultural Practices

Training and Pruning

Training : The ber plants start bearing within 2-3 year of their planting in the orchard. In the fourth year, the trees bear commercial crop. Hence, the ber plants should be properly trained during the first 2-3 years, to build up a strong framework. The young budlings should be given support with a bamboo stake to avoid the breakage of the bud-union and to support the main stem. The commercial varieties are spreading in nature, therefore, staking is absolutely essential during the first two years to train the tree properly. The plant will make vigorous growth during the first year. Many secondary branches will emerge from the main stem. All the branches upto 75 cm from the ground level should be removed. Four or five laterals which are most favourably located around the main stem should be selected to make the proper framework. At the end of the year, the main stem should be headed back to some outgrowing laterals.

Pruning : The pruning of ber trees is highly desirable to maintain their vigour and productivity as well as to improve fruit size and quality. Pruning also saves the fruit from being affected by the powdery mildew disease and strong winds. The ber tree remains young upto 30 years, if proper pruning is done regularly. Ber fruit is borne in the axils of leaves on the younge growing shoots of the current year. Hence, a regular annual pruning is essential to induce a good and healthy growth which will provide a maximum fruit bearing area on the trees as well as to improve the fruit size and quality. In unpruned ber tree, the canopies of the trees get un-necessarily enlarged, the growth and branchlets become weak and both fruit size and quality gets impaired. Ultimately, such trees become economically unproductive besides occupying large orchard space. Some thinning out of the branches of ber trees is also necessary to avoid too much crowding so as to admit adequate sunlight and facilitate proper aeration.

Ber pruning experiments have shown that the light pruning, i.e. heading back of 25 per cent of the previous year's growth (branchlets, shoots, etc) is desirable to obtain heavy yield, good fruit size and better quality. The lower branches should be pruned suitably to prevent them from spreading on the ground. The diseased, broken and intercrossing branches should also be thinned out. Severe pruning after every four-five years, is recommended. The ber trees shed their leaves and enter into dormancy by the end of May. The best time for taking up pruning would, therefore, be end-May or beginning of June.

Irrigation

Irrigation is essential during the development of fruit, i.e. from October to February at intervals of 3 or 4 weeks depending upon the weather. Trees will continue to bear even if no irrigation is applied during this period but the yield is substantially reduced because of heavy fruit drops and smaller size of the remaining fruit. The quality of fruit is also very poor. It has been observed that the fruit become large and their quality is improved the fruit shedding is very much minimised if irrigation is applied during fruit development period. Irrigation should be stopped in March as fruits on the branches lying on the ground get damaged and their ripening is delayed.

The harvesting of fruit is over in April and they become dormant in May-June and shed their leaves. They need little or no irrigation during this period. If irrigation is applied during the dormant period, the trees would continue to put fourth growth haphazardly which is not desirable. Under Panjab conditions there are sufficient rains during July to September when the tree produce the maximum fresh growth. During the second half of September and in October the trees come into flowering. At this time, light irrigation should be given.

Manuring and Fertilization

Proper nutrition of ber tree is necessary to get good crop over the years. The fruit becomes large and attractive and get decent price in the market. 20 kg farmyard manure and 100 g nitrogen (400 g CAN) is recommended for one year old ber tree. Similar amount of farmyard manure and nitrogen should be increased every year up to the age of five years. The quantity of farmyard manure and nitrogen should be stabilized at 100 kg and 500 g (2 kg CAN), respectively, after the age of five years.

Farmyard manure should be supplied in May-June. Half of the CAN may be applied during rainy season (July-August) and the other half at the time of fruit-set (October-November). The fertilizer should be evenly spread in the basins of trees upto the periphery. After adding the fertilizer, light hoeing with spade or khurpa should be given to the basins to mix it thoroughly with the soil.

Intercropping

The ber tree begins to bear after one year of its planting in the field. To develop the tree properly, it is advisable that no fruit should be taken at least for the first two-three years. Intercropping can be successfully practised on the vacant land in the young orchard during the first four years. Only leguminous crops of short stature like gram, moong and mash can be grown to get some income from the land in these initial years. These crops also enrich the soil by fixing nitrogen. The other exhaustive and tall-growing crops should not be grown in the ber orchard as they deplete the soil of its nutrients to a greater extent and compete for light with the trees.

Manures and fertilizers, irrigation and plant protection measures should be given seperately to the fruit trees and intercrops according to their needs.

Weed Control

Pre-emergence application of Hexuron 80 WP (diuron) at 1.2 kg/acre can be made during the first fortnight of August when field is free from growing weeds and stubbles. Glycel 41 Sl (glyphosate) at 1.2 litres/acre or gramoxone 24 WSC (paraquat) at 1.2 litres/acre as post-emergence should be sprayed when the weeds are growing actively preferably before weeds flower and attain a height of 15-20 cm. Dissolve the herbicide in 200 litres of water per acre to give complete coverage of weeds/field. Spray Glycel and Gramoxone during the calm day to avoid spray drift to the foilage of the fruit trees.

Control of Fruit Drop : Spray application of 20-30 ppm Naphthalene acetic acid once in the second week of October and again in the second week of November, check 11 per cent fruit drop in Sanaur-5 snd 10 per cent in Kaithli cultivers of ber.

Harvesting and Fruit Handling

Harvesting : The ber tree grows quickly and and the first crop can be harvested within 2-3 years of planting. The fruit itself requires about 22-26 weeks to mature after fruit-setting. The peak season of harvesting in north India is in mid-March to mid-April but some early varieties may ripen by end-February. This period being a slack season for other kinds of fruits, ber sells readily at remunrative prices.

The fruit should always be picked at the right stage of maturity, i.e. when it is neither under-ripe nor over-ripe. It should be picked when it has acquired normal size and characteristic colour of the variety, e.g. golden yellow colour in Umran. Normally four to five pickings have to be made as all the fruits on the tree do not ripen at the same time. In no case, the fruit should be allowed to become over-ripe on the trees, as they deteriorate in taste and quality and thus fetch lower price in the market.

Insect-pests

Fruit-fly (Carpomyia vasuviana costa) : It causes great damage to ber fruits. The larvae feed inside the fruits and render them unfit for human consumption. To control the pest, pick and destroy the infested fruits and spray 500 ml of Roger 30 EC (Dimethoate) in 300 litres of water during February-March, care being taken that sprayings are stopped atleast 15 days before fruit harvest.

Leaf-eating caterpiller (Porthmologa paraclina Meyrick) and ber beetle (Adoretus pallens Harold) : The plantation should be watched carefully during rainy season regarding the attack of these insect-pests. Leaf-eating caterpillars feed on leaves and cause huge damage. To control these, spray with 750 g Sevin 50 per cent (Carbaryl) in 250 litres of water as soon as the damage is noticed.

Lac insect (Laccifer lacca): This insect also causes serious damage by sucking the sap from the twigs which usually dry up. To control this insects, remove and destroy the infested dry twigs. Spray the trees with 250 ml of Rogor 30 EC or 100 ml of Dimecron 85 WC (Phosphamidon) in 250 litres of water in April and again in September.

Diseases

Powdery mildew : The disease is caused by Microsphaera alphitoides f.sp. Zizyphi and appears from September to December. It has become a big meance to ber orchards in north India. If not checked in time, the disease can wipe out the entire crop. Young developing leaves and fruits are covered with withish powdery mass of the causal fungus. The disease cause premature defoliation and heavy fruit-drop. Affected fruits remain small and become cankered and disfigured. Sometimes the attack is so sereve that the entire crop is lost either through drop or rendered unmarketable, thus causing heavy economic losses to the growers. The disease can be controlled by 3-4 spray of 0.05 per cent Karathane 40 EC (50-80 ml in 100 litres of water) or 0.25 per cent wettable sulpher (250 g in 100 litres of water). First spray must be given before flowering (first fortnight of September), second spray after fruit-set in early October and the third in the end of October. Another spray can be given if need arises.

Leaf spots : Two leaf spots of ber are very common in ber growing regions, i.e. 'Phoma Leaf Spot' caused by Phoma macrostoma Mont. and 'Black Mould of Leaf' caused by Isariopsis indica. Both diseases are caused by different fungi, while the Phoma leaf spot appears on the upper surface, the black mould make its appearence only on the lower surface of leaves. In case of Phoma leaf spot symptoms appear when the leaves have fully expanded, in the Black mould case the sysmptoms can appear even on young leaves. Phoma leaf spot appears with grey centre, yellow margin and dark fungal growth on the mid-rib, main vein, petiols and the leaves. Black mould spot appears as small circular, small finger-like projections like softy tufts. Both the leaf spots of ber can be controlled by spraying the following fungicides as given below :

First spray : Bordeaux mixture 2 : 2 : 250 or with 0.3% copper oxychloride 50% (300 g in 100 litres of water) should be sprayed both on upper and lower surface of leaves with the appearance of disease in August or when the leaves have expanded.

Second spray : This spray should be given after 14 days of the first spray with 0.2 per cent Dithane M-45 WP 75% (200 g per litres of water) both on upper and lower surface of leaves.

First and second spray should be repeated alternatively at 14 days interval till the fruits are fit for marketiing. Thereafter sprayings are stopped a week before harvesting.

JAMUN

Climate

Jamun is successfully grown under tropical and subtropical climate. It also occurs in the lower range of the Himalayas upto an elevation of 1300 metres and in the Kumaon hills upto 1600 metres above sea level. It is widely grown in the larger parts of India from the Indo-Gangetic plains in the North to Tamil Nadu in the South. It is one of the most hardy fruit crop and can be easily grown in neglected and marshy areas where other fruit plants can not be grown successfully. Jamun is somewhat more susceptible to cold and drought than the seedling mango trees. It requires dry weather at the time of flowering and fruiting. For ripening of fruit and proper development of its size, colour and taste, early rains are considered very beneficial. The fruits show remarkable improvement in these respects after the very first shower of rain.

Soil

Jamun tree is exact in its soil requirements. The tree requires deep, loam and well-drained soil for its optimum growth and good fruiting. Its cultivation should be avoided in very heavy or light soils.

Choice of Varieties

Ra Jamun : It produces big sized fruit with average lenght of 2.5 - 3.5 cm and of diameter 1.2 - 2.0 cm. Fruits are oblong in shape, deep purple or bluish black in colour at fully ripe stage. The pulp colour of ripe fruit is purple pink and the fruit is juicy and sweet. The stone is small in size. It ripens in the month of June-July. The variety is very common among the people.

Small sized Jamun : It is a late maturing variety. The average lenght of fruit is 1.5 - 2.0 cm and diameter is 1 - 1.5 cm. The fruit is slightly round in shape, deep purple or blackish in colour at full ripe stage. The colour of the pulp is purple, less in juice, weight and sweetness of pulp in comparison of that of 'Ra Jamun'. The stone is very large. Fruits ripen in the month of August.

Propagation and Rootstock

Though lot of jamun plantation is seen on road side or scattered plantation at farmer's field, yet no single well established orchard is reported in North India. Also no systematic work has been done on its propagation.

The most common method of jamun propagation is by seed. Seedling plant bears fruit of variable size and quality. For improved and selected true-to-true, vegetative methods of propagation like inarching and air-layering have been advocated.

Sexual Propagation : Seeds are sown fresh in flat nursery beds during July when this fruit ripens. The seeds should be sown at a distance of 15 cm in rows which are 25 - 30 cm apart. The seedlings can also be raised in polythene bags of 22.5 - 30 cm size. The bags should be filled with a mixture of soil and farmyard manure in equal proportion. To drain out excess water the polythene bags should be pricked from all sides before filling the mixture. It has been seen that more than one seedling comes out from a single seed. These seedlings are separated in different bags when they are about two weeks old.

Plants grown from seed become transplantable during next spring season. But it is advisable to keep them in nursery upto next rainy season which is the best time of its plantation.

Vegetative Propagation

T-budding and patch budding : According to the recent research conducted at Punjab Agricultural University has revealed that jamun can be best propagated through T-budding as well as through patch buddding. The seedling of jamun is used as a rootstock. The per cent success is higher in T-budding (70%) than patch budding (60%). The best time for budding in both the cases is either during February-March or during August-September. However, the success during August-September is higher.

Inarching : The rootstock used for propagating jamun is the jamun seedlings. For raising the rootstock, seeds are collected from healthy, vigorously growing and high yielding jamun trees. Seedlings are raised either in bed or in pots singly. In the month of June-July one or two year old rootstocks are inarched with the matching thickness of scion. Rootstocks are watered if necessary till the grafts are separated from the parent tree. The union will complete in a period of about six weeks.

Veneer grafting : Veneer grafting gives 31 per cent success when one year old seedlings are used as rootstocks. The shoots are taken from spring flush and the method is employed in the month of July.

Planting Operation

Jamun can be transplanted during spring (February-March) or during monsoon (August-September). However, the later season of planting is considered better because the plants easily get established during the rainy season. The plants are transplanted with earth ball and are given irrigation till they get established. The size of the pit should be 1 × 1 × 1 m and these should be filled with a mixture of surface soil, silt and well rotten farmyard manure. The jamun is planted at the distance of 10 - 12 metres in square system, thus accommodates 105 - 75 plants per hectare.

Flowering and Fruiting

The flowering in jamun starts in the first week of March and continues up to the middle of April. The trees are in full bloom in the second week of April. The inflorescence in jamun is generally borne in the axils of leaves on branchlet. The flowers are hermaphrodite, light yellow in colour. The maximum anthesis and dehiscence were recorded between 10 A.M. and 12 Noon. The pollen fertility was higher in the beginning of the season. The maximum receptivity of stigma was observed one day after anthesis.

Jamun is a cross-pollinated fruit. The pollination is done by honeybees, house flies and wind. The maximum fruit set i.e. 32.6 - 36.0% was obtained when pollination was done one day after anthesis. Thereafter, a sharp decline was observed in fruit set.

Three distinct phases of fruit growth in jamun are recorded. During the first phase (15 - 52 days after fruit set), the rate of growth was slow. In the second phase (52 - 58 days after fruit set), the rate of development was quite rapid and the third and last phase (58 - 60 days after fruit set) comprised comparatively slow growth with little addition of the fruit weight. The lenght and diameter of fruit showed a continues increase with advancement of maturity. The colour of jamun fruit changed from dark green at fruit set to light reddish colour at partial ripening and dark or bright purple at full ripe stage. The fruit took 63 days for complete ripening from fruit set. The ripe jamun had 76 per cent edible portion and 3 : 1 : 1 pulp to seed ratio. TSS and sugars followed an increasing trend, while tannin content followed a decreasing trend during growth and development.

Orchard Cultural Practices

Irrigation

During the initial year, the jamun plants required 8 - 10 irrigations in a year while bearing trees (full grown) required 4 - 6 irrigations during the summer months of May and June when fruits ripen. In the remaining period of the year, irrigation may be given when there are no rains or dry spell persists. During winter month, irrigation proves useful as it protects the plants from frost injury.

Intercropping

To supplement the income from pre-bearing period of jamun, intercropping should be practised judiciously. Intercropping also improved fertility of the soil. Fruit crops like peach, pulm, guava, kinnow, kagzi lime, phalsa and papaya can be grown as filler trees. Such filler trees can be uprooted when the jamun trees starts bearing commercial crop. The leguminous crop like gram, peas, moong and mash can be grown successfully. In addition, intercrops of vegetables near established market may be taken with cauliflower, cabbage, knol, khol, radish, brinjal, turnip, carrot etc.

Manuring and Fertilization

During the pre-bearing period of jamun, a dose of 20 kg well rotten farmyard manure should be applied. To the bearing tree 80 kg FYM per tree should be supplied annually for proper growth and fruiting. Sometimes in highly fertile soils, the plants produce profuse vegetative growth and fruiting is delayed. Under such conditions, the manures should not be given and irrigation should also be given sparingly and withheld in September-October and again in February-March. This procedure will prove beneficial in fruit bud formation, flowering and fruit setting. Ringing and root pruning are also helpful.

Training and Pruning

Jamun plants should be trained according to the modified leader system. Regular pruning is not required in jamun plants. However, in later years, the dry twigs and crossed branches are removed. While training the plant, the framework of branches is allowed to develop above 60 - 100 cm from the ground level.

Flower and Fruit Drop

In jamun, the flower and fruit drop start just after opening of flowers and continue up to maturity. About 65 per cent flower and fruit drop in the first five weeks and since then a maxium of 19 - 21 per cent flowers and fruits drop off up to maturity. Only 12 - 15 per cent flowers reach maturity. The flower and fruit drop are found at 3 stages. The first drop takes place during bloom or shortly there after. This proves to be the heaviest drop as about 52% of the flowers drop off after 4 weeks from flowering. The second drop starts about 35 - 40 days of full bloom and apparently there is no difference between the developing and aborting fruits. The third drop takes place after 42 - 50 days of full bloom and continues till 15th July.

Control : The extent of flower and fruit drop in jamun may be reduced by two sprays of 60 ppm GA3, one at full bloom and the other 15 days after initial setting of fruit.

Harvesting and Fruit Handling

The grafted jamun starts bearing after 6 - 7 years while the seedling one after 8 - 10 years of planting. The fruit ripen in the month of June-July. The main characteristic of ripen fruit at full size is deep purple or black colour. The jamun fruit is non-climacteric in nature. The fruit should be picked immediately when it is ripe, because it can not be retained on the tree in ripe stage. The ripe fruits are picked singly by hand and in all cases care should be taken to avoid all possible damage to fruits. For harvesting, the picker climbs the tree with bags of cotton slung on the shoulder. The fruits of jamun is generally harvested daily and sent t market on the same day. The fruit is highly perishable and can be kept in good condition for about 2 - 3 days under ordinary conditions.

The average yield of fruit from a full grown seedling jamun tree is about 80 - 100 kg and from a grafted one 60 - 70 kg per year. The storage life of jamun fruit is 6 days at room temperature and 3 weeks at low temperature (9°C and RH 85 - 90%) when pre-cooled fruits are kept in perforated polythene bags.

Insects-pests

White-fly (Dialeurodes eugeniae)

It damages the tree in all parts of India. Sometimes the fruits of jamun get wormy due to attack of fruit fly. Pests can be controlled by maintaining sanitary situation in the orchard. Pick up the affected fruits and burry them deep in soil. The area under the tree should be dug so that the maggots in the affected fruits and the pupae hibernating in the soil may be destroyed.

Leaf eating caterpillar (Carea subtilis)

It is reported to damage the plant in South India at Coimbatore. The insect infest the leaves and may defoliate the trees. The pest can be controlled by spraying 625 ml Roger in 500 litres of water.

Squirrels, Parrots, Crows and Birds

The jamun fruits are also damaged by these squirrels, parrots, crows and birds. For keeping them away, beating of drum or flinging small dry earthern balls through a sling is useful.

Diseases

Leaf spot and fruit rot

The disease caused due to Glomerella cingulata. Affected leaf shows scattered spots, light brown or reddish brown in colour. The affected fruits rot and shrivel. The disease can be controlled with fungicide like Dithane Z - 78 at 0.02% or Bordeaux mixture (2 : 2 : 250).

REFERENCES

Vishal Nath, Dinesh Kumar and V. Pandey, *Fruits for the Future*, Vol. I. *Well Versed Arid and Semi Arid Fruits*, Delhi, Satish Serial Pub., 2008.

Utpal Banerjee, *A Handbook of Practical Horticulture*, Jaipur, Mangal Deep, 2002.

Saroj, P.L. Vashishtha, B.B. Dhandar, D.G., *Advances in Arid Horticulture : Vol. I : Present Status*, Lucknow, International Book Distributing Co., 2004,

Atul Chandra, *Arid Fruit Research*, Anju Chandra and I.C. Gupta. 1994.

Sharma, R.R., *Fruit Production : Problems and Solutions*, Lucknow, International Book Distributing Co., 2006

4

Production of Temperate Fruits

APPLE

Apple (*Pyrus malus L.*) occupies nearly 12,141 hectares, mostly in temperate regions of Punjab, Uttar Pradesh, Himachal Pradesh and Kashmir and to a small extent in the Nilgiris.

There are more than 7,500 known cultivars of apples. Different cultivars are available for temperate and subtropical climates. Reputedly the world's biggest collection of apple cultivars is housed at the National Fruit Collection in England. Most of these cultivars are bred for eating fresh (dessert apples), though some are cultivated specifically for cooking (cooking apples) or producing cider. Cider apples are typically too tart and astringent to eat fresh, but they give the beverage a rich flavour that dessert apples cannot.

Commercially popular apple cultivars are soft but crisp. Other desired qualities in modern commercial apple breeding are a colourful skin, absence of russeting, ease of shipping, lengthy storage ability, high yields, disease resistance, typical "Red Delicious" apple shape, long stem (to allow pesticides to penetrate the top of the fruit), and popular flavour. Modern apples are generally sweeter than older cultivars, as popular tastes in apples have varied over time. Most North Americans and Europeans favour sweet, subacid apples, but tart apples have a strong minority following. Extremely sweet apples with barely any acid flavour are popular in Asia and especially India. As an example, the U.S. state of Washington made its reputation for apple growing on Red Delicious. In recent years, many apple connoisseurs have come to regard the Red Delicious as inferior to cultivars such as Fuji and Gala due to its merely mild flavour and insufficiently firm texture.

Old cultivars are often oddly shaped, russeted, and have a variety of textures and colours. Some find them to have a better flavour than modern cultivators, but may have other problems which make them commercially unviable, such as low yield, liability to disease, or poor tolerance for storage or transport.

A few old cultivars are still produced on a large scale, but many have been kept alive by home gardeners and farmers that sell directly to local markets. Many unusual and locally important cultivars with their own unique taste and appearance exist; apple conservation campaigns have sprung up around the world to preserve such local cultivars from extinction.

Varieties

Apple varieties fall into two categories: diploids and triploids. Diploids have plenty of good pollen and are self-fruitful. Triploids are self-unfruitful and become productive only when pollinated by using suitable pollenizer varieties. Even self-fruitful varieties have to be interplanted to get commercial crops through cross-pollination.

Varieties selected for interplanting should sufficiently overlap in their blossoming periods. Important varieties are listed below:

— *Himachal Pradesh.* 'Red Delicious', 'Golden Delicious', 'Worester Pearmain', 'Newton Wonder' (all diploids),

'Cox's Orange Pippen' (triploid), 'King of Pippens' (No. 13), 'Starking (Royal) Delicious' and 'Richard'.

— *Kashmir Valley*. 'Red Delicious' (diploid), 'Baldwin' (triploid), 'Ambri Kashmiri', 'White Dotted Red' and 'Blood Red'.

— *Simla Hills*. 'Beauty of Bath' (triploid), 'Red Delicious', 'Jonathon', 'Rome Beauty' (all diploids), 'Early Shanburry', 'Red Astrachan', 'Red Sudeley', 'Stayman Winesap', 'Winter Banana' and 'Yellow Newton'.

— *Kumaon Hills*. 'James Grieve', 'Jonathon', 'Rome Beauty' (all diploids), 'Blenheim Orange Pippen', 'Delicious', 'Early Shanburry', 'Golden Pippen', 'King of Pippens', 'Rhymer' and 'Winter Banana'.

— *Kulu Valley*. 'Ben Davis', 'Red Delicious', 'Golden Delicious' (all diploids), 'Cox's Orange Tippen', 'Blenheim Orange', 'Baldwin' (all triploids), 'Red Astrachan', 'King of Pippens', 'Yellow Newton' and 'Granny Smith'.

— Nilgiri. 'Rome Beauty' (diploid) and 'Irish Peach'.

Propagation and Planting

Propagated mainly by shield-budding, bench-grafting and tongue-grafting on seedlings raised from seed. Use M. IX dwarfing stock and 7.5 to 9 metres if on seedling stock. Standardized clonal rootstocks of the Malling-Merton series are recommended where woolly aphis is serious.

APRICOT

Varieties

The cultivated varieties of apricot are mainly exotic and they grow successfully at varying elevations. The following varieties are recommended : 'Shipley Early', 'Kaisha', 'New Castle', 'St. Ambroise', and 'Royal'. All these varieties are self-fruitful.

Propagation and Planting

Apricot is propagated by shield-budding on wild apricot stock, i.e. zardalu. Peach stock may also be used. Plant one-year-old grafts in autumn 6 to 7.5 metres apart.

Pruning

Apricot grafts usually have numerous lateral branches unlike the straight whips of apple, pear and cherry. If the laterals developed in the nursery are not properly spaced, cut off the main stem while planting, about 50 to 75 cm above the ground level to promote the growth of new laterals.

During the first summer, remove all unwanted laterals, leaving three to five well-placed branches to form the framework. Head-back the scaffold branches breast-high next winter to get secondary scaffold branches. Leave the upper branches longer than the lower ones, as the latter grow faster and crowd out the upper branches.

By the end of the second growing season, the tree produces a large number of laterals on the scaffold branches and trunk, which should all be removed, except a few short growths (7 cm to 12 cm long) on the trunk and the main branches. Retain only five to seven secondary scaffolds. In subsequent years, thin only the branches which are either crossing or crowding one another.

This practice admits light into the centre and encourages the growth of spurs. The pruning of old trees should aim at producing new spurs to replace those broken during picking. The kind and the amount of pruning depend upon the bearing habit of the variety. Light to moderate thinning of branches and the shortening of new wood back to the laterals is the usual practice. If new growth is less than 40 to 80 cm each year, resort to severe pruning.

Harvesting

The fruit should be picked when it is still hard, but has attained the proper colour. For drying, the fruit is harvested by hand-picking when it is fully ripe.

CHERRY

Cherries (*Prunus avium L.*) are of two types : sweet used for desert, and sour used for cooking, grown mainly in the Simla Hills, the Kulu Valley and Kashmir at elevations above 1,500 metres.

Varieties

Selected varieties of proven merit are : 'Early Rivers', 'Governor Wood', 'Bigarreau de Schreken', 'Elton','Bedford Prolific', 'White Bigarreau', 'Monstrueuse de Mezel', 'Bigarreau Napoleaon', 'Emperor Frencis' and 'Late Black Bigarreau'. It is desirable to choose varieties that will ripen in succession in order to obtain the crops over a longer period. A large number of varieties are self-unfruitful and do not set fruits with their own pollen. As they are also cross-incompatible, only the compatible varieties, whose period of flowering overlaps to effect cross-pollination, should be interplanted to get commercial crops.

Propagation and Planting

The plants are propagated by whip or tongue grafting on seedlings of wild cherry stock, called paja. Grafts are ready for transplanting in two years. Sometimes, the rootstock plants are planted in permanent positions in the orchard and grafted in situ.

As cherry-trees are generally affected by frost, the site selected for planting should be such that the sun reaches the trees gradually. The distance between the trees varies from 9 to 12 metres, depending upon the variety. The trees should be properly staked after planting.

Pruning

Cherry-trees grow into shape without much pruning. Crowded branches should be thinned out and dead-wood removed in the dormant season. The pruned cuts should be painted with tar.

Manuring

Cherry orchards are best put under grass which is grazed by sheep. In addition to sheep manure, phosphate manures are applied to obtain a good growth of clovers. A dressing of fertilizers to supply 75 to 100 kg of N, 55 to 90 kg of P_2O_5 and 110 to 165 kg of K_2O per hectare may be recommended.

PEACH

The area under peach (*Prunus persica (L.)* Stokes) is very small and is mainly located in the Himalayas at various elevations.

Varieties

Some of the promising varieties are 'Early Beatrice', 'Alexander', 'Early Rivers', 'Duke of York', 'Peregrime', 'Noblesse', 'Late Devonian', 'Elberta', 'J.H. Hale', and 'Triumph'. Except 'J.H. Hale', all other varieties are self-fruitful and set good crops without cross-pollination.

Propagation and Planting

Propagation is done by budding on seedling peach. One-year-old grafts are planted 6 to 8 metres apart in early spring. Immediately after planting, the trees are white-washed to protect the bark from the sun.

Pruning

At the time of planting, the stem is cut to about 0.6 metre from the ground and three to four branches are allowed to develop, distributed round the main stem. All other shoots that grow during the first summer are removed. During the first dormant season, two well-spaced secondary branches on each main branch are selected and the main branch is cut close t the secondary branches.

During the second summer, water-sprouts, if any are removed. At the time of second pruning in winter, secondary branches are not cut, except to regulate the shape of the tree. In pruning, cut always to the outside buds to encourage a spreading shape.

In the case of bearing trees, annual pruning is necessary to maintain the centre open. Two- to three-year-old branches may be cut back to the outward-pointing side branches to encourage a spreading growth. Shorten and thin outside branches to stimulate the growth of new fruiting wood every year. A satisfactory annual growth should be 45 to 50 cm long.

Fruit-buds are borne laterally upon one-year-old wood and on short spur-like twigs. Ordinarily, they develop two fruit-buds and a leaf-bud at one node. The fruit-buds are usually located from the middle of the shoot upwards. In cutting away branches, the position of the fruit-buds should be taken into consideration.

Culture

A peach orchard should be regularly cultivated. Ploughing, which should not be deeper than 10 cm, is generally done in winter. A suitable cover or green-manure crop may be sown in the rainy season after the fruits are picked and ploughed-under during winter. A dose of fertilizers to supply 55 to 65 kg of N, 55 to 65 kg of P and 110 to 135 kg of K per hectare may be applied to the bearing trees in spring. Immediately after the natural fruit-drop in May and June, the fruits should be thinned out so as to have them 10 to 15 cm apart.

Harvesting and Marketing

Peaches are picked when they are still hard, as they can ripen well during storage or in transit.

PEAR

Pear (*Pyrus communis L.*) is grown mainly in the hills at elevations ranging from 1,500 to 2,500 metres. Its cultivation is rather restricted, mainly because the fruit does not store well.

Varieties

The following varieties are recommended : 'William Bon Christien' (Bartlett), 'Clapp'soil Favourits', 'Thimpsons', 'Doyenne du Comice', 'Easter Beurre', 'Winter Nalis', 'Conference', 'Dr, Jules Guyot', 'Marie Louise d'Uccle', Baggugosha (Citron des carmes) and Emile d'Heyst. Baggugosha can also be grown in the submontane tracts, but there its quality is poor. Nashpati is another variety that is grown successfully in the plains.

Pollination

Most of the pear varieties are self-unfruitful and the planting of pollenizer varieties is advocated. Nashpati is a self-fruitful variety.

Propagation and Planting

Pear is propagated by shield-budding which is done in June-July. The stocks are raised either from the seeds of a commercial variety or from those of wild pear, shegal (*Pyrus pashia*). To produce dwarf trees, quince 'C' stock is employed. Some varieties are not compatible with quince. They are propagated by double-working, using as intermediate a pear variety which can successfully be grown on the quince stock.

One-year-old grafts are planted in autumn. Those propagated on the quince stock are planted one to one-and-a-half metres apart, if they are trained as cordons. Those trained as pyramids are planted 3.5 to 4.5 m apart.

Pruning and Training

Pears on the pear stock make vigorous growth and develop into large trees. They remain dwarf on the quince stock when they are trained into different forms. For the pyramid form, cut the graft while planting at about three-fourths of a metre above the ground. Next winter, prune the leader to about 25 cm and the laterals to about 20 cm to the outward-pointing buds. In the second summer, all the branch leaders and laterals should be pruned to five or six leaves from the clusters, allowing the central leader to grow unchecked. In the third winter, the central leader is cut back to about one-third its length, but the branch leaders and laterals are not pruned. In the third summer, the branch leaders and the laterals, except the central leader, are again cut back to five or six leaves as in the previous summer. In the fourth winter, the central leader is again cut back to one-third its length. By following this procedure, a pear-tree on the quince stock would start flowering in fourth year. The bearing pear-trees are pruned as in the case of apples.

Fruit-thinning

As a rule, less thinning is required in the case of pears than in the case of apples. One fruit per cluster in the case of prolific varieties and one to two fruits per cluster in others may be retained after thinning.

Manuring

The method of manuring and the time of its application are the same as for apples. The amount of nitrogen to be applied should, however, be a little more than in the case of apples.

Harvesting and Marketing

'Bartlett' is picked when still green and hard. The early varieties are packed without storing, whereas the late varieties require storing to develop full flavour. The fruit should be size-graded before storing.

PERSIMMON

Persimmon (*Diospyros kaki L.f.*) is grown in the Kulu Valley at elevations ranging from 900 to 1,500 kilometres. The promising varieties are 'Fuy', 'Hachiya' and 'Hyakume'. Several good varieties, such as 'Dai Dai', 'Maru' and 'Tenanshi', are also being grown successfully at the Pomological Station, Conoor in the Nilgiris in southern India. The tree is propagated by grafting (whip and tongue method) on seedlings of Diospyros lotus and D. virginiana. The grafts are planted in winter, 6.5 to 7.5 m apart.

The trees are headed back one or two metres above the ground at the planting time. Four to five shoots are allowed to grow round the stem to avoid narrow crotches and to develop a well-balanced head. There is no further pruning after this. Dead, broken and interfering branches are removed every year.

The fruit is picked when it has attained a yellowish or reddish colour, characteristic of the variety, when still hard. It is clipped from the tree, keeping intact the calyx and a short piece of the stem. It is wrapped up in tissue-paper and packed in a two-layer box for transport. With astringent varieties, the fruit has to be cured before it is fit for eating out of hand. The simplest method is to place the fruits in a closed chamber with other ripening fruits such as pears and tomatoes.

PLUM

Plum (*Prunus domestica (L.) Stokes*) is grown mainly in the Himalayas where the following varieties have been successfully grown : 'Grand Duke', 'Early Transparent Gage', 'Victoria', 'Santa Rosa', 'Wickson', 'Beauty' and 'Kelsey'. In the south, in the Nigiris and Kodaikanal in the Tamil Nadu stae, several choice varieties of the Japenese plum (*P. salicina*) are grown. The more important of these are 'Rubio', 'Alu Bokhara', 'Gaviota', 'Shiro', 'Combination' and 'Hale'. All varieties except 'Beauty', 'Santa Rosa', 'Gaviota', 'Rubio', 'Alu Bokhara' and 'Hale' which are self-fruitful, requires cross-pollination from other varieties. Plums are usually propagated by shield-budding on wild apricot or common peach stock. Planting, spacing, cultivation and fertilization are the same as for peach.

Pruning

Cut back the top to about 60 cm at planting time. Select three to five scaffold branches situated spirally round the stem, equidistant from one another, and remove the unwanted ones. At the time of first winter pruning, the main branches are headed back. All growth, except the main and secondary branches, is removed during the year. At the second winter pruning, crossing and other undesirable branches are removed. In the case of varieties having a tendency for upward growth, heading should be done to outward-pointing buds to make them more spreading. Subsequent pruning is carried out every year on similar lines. The pruning should be light as far as possible. The bearing trees are pruned to secure a balance between vegetative growth and fruiting.

Thinning

Thinning should be carried out after the natural fruti-drop in April and May but before the hardening of the pits.

Harvesting and Marketing

For transporting, the fruits are picked a few days in advance of full maturity. The change of colour for each variety determines its stage of maturity. The fruits are required to be harvested in three or four pickings.

STRAWBERRY

The cultivated varieties of strawberry (Fragaria spp. grown in India are all imported. The following are recommended : 'Laxton's Latest', 'Royal Sovereign', 'Early Cambridge', 'Huxley Giant', 'Penomenol' and 'Robinson'.

Propagation and Planting

Maiden palnts (runners) that have not borne any crop are used for planting. The planting-distance is half a metre between plants and three-fourths to one metre between rows. Runners with a good root-system are used to set a new palntation. Transplanting is done in March-April in the hills and in January-February in the plains.

Culture

Prepare the land by ploughing deep, followed by harrowing. Add bulky organic manures. Keep down weeds by light hoeing and runners, as and when they form. Manuring is done in winter. When plants blossom in spring, bed the plantation with straw to keep fruits off the soil. After fruiting, remove the straw and weeds, and cut off all runners. Continue hoeing. Rotate strawberry with vegetables every three years.

REFERENCES

Chundawat B.S., and Sen, N.L., *Principles of Fruit Culture*, Udaipur, Agrotech, 2002.

Donald L. Plucknett and Howard B. Sprague, *Detecting Mineral Nutrient Deficiencies in Tropical and Temperate Crops* Reprint. 1991.

Jitendra Singh, *Spices and Plantation Crops,* Jaipur, Aavishkar Publishers, 2008.

Kishore, D.K., Satish K. Sharma and Kallol K. Pramanick, *Temperate Horticulture : Current Scenario,* New Delhi, New India, 2006.

5

Production of Vegetables

Vegetables constitute an important item of human diet. In the context of alleviating protein malnutrition in India, efforts are under way to enrich cereals. To supplement them vegetables, being short-duration crops, can be produced in succession on the same plot and all the family labour of the vegetable-grower can be usefully employed throughout the year. The daily minimum requirement of vegetables, according to a dietician, is 284 g per head, i.e. about 20 per cent of the daily requirement of the total food of an adult. This requirement is more in the case of a vegetable diet.

The present production and consumption of vegetables in the country are very inadequate, being only about one-fourth to one-third of the requirement. In order to improve the quality of the diet of the people, it is essential that the production of vegetables should be increased considerably. This object can be achieved by increasing the present area under vegetables and also by increasing the yield per unit area by adopting better agricultural techniques.

Like any other crop, seed is an important factor governing the production of vegetables. Seed production is technical job in the case of vegetables , unlike in the case of most other crops. The production techniques are also different in biennial vegetables, including onion, cabbage, beet, etc., and specific environmental conditions are required for producing their seeds. The growers depend mostly on outside agencies for meeting their requirements of seeds. Seed production and trade have not developed in this country on any scientific lines and that the approach is not systematic. That is why the quality of the vegetable seeds has not been what it should be. The National Seeds Corporation, a Government of India undertaking, has recently entered the vegetable seed industry. Some private seed nurseries have also employed qualified technical personnel. A Central Seed Act has also been passed by the Parliament. These are all sure steps to streamline seed production, inspection and certification.

The growing psychology of the farmers in recognizing the need for quality seeds augurs well for the production of vegetables. The quality seed production will not only help to increase the production within the country but will also open up a big source of potential foreign exchanges\ by exporting vegetables seeds to our neighbouring countries in South-East Asia and Africa.

TYPES OF VEGETABLE GARDENS

Vegetable gardens can be classified into six different types according to the purpose for which they have been developed. These are home-gardens or kitchen-gardens, market-gardens, truck-gardens, gardens for processing, gardens for vegetable-forcing, and gardens for seed production. The layout of a home-garden will differ from individual to individual. However, broadly, a city home-gardener will follow a very intensive method of vegetable-growing compared with that followed by a home-gardener in a village.

A market garden produces vegetables for the local market. Most of such types of gardens are located within 15 to 20 km from a city. The cropping pattern depends on the demands of the local market. A truck-garden produces selected crops in a relatively large quantity for distant markets. It generally follows a more extensive method of cultivation than the market-garden. The location is determined by soil and climatic factors suitable for raising particular crops. A vegetable garden for processing develops around the processing factories and is mainly responsible for supplying vegetables to the factories regularly. This type of garden grows particular varieties suitable for canning, dehydration or freezing. Vegetable-forcing is concerned with the production of vegetables out of the normal season. The commonest forcing structures are glass and plastic houses.

Vegetable seed production is rather a specialized type of vegetable-growing. A thorough knowledge of a vegetable crop in respect of its growth habits, mode of pollination, proper isolation distance, etc. are of prime importance in the production of quality seed. The handling of the seed-crop, its curing, threshing, cleaning, grading, packing and storage need specialized knowledge. A vegetable garden for seed production is , therefore, considered a special type of garden.

CLASSIFICATION OF VEGETABLES

There are more than fifty important vegetables. If the growing of each is dealt with in details, it will lead to much repetition. It is, therefore, desirable to classify the vegetables into certain groups. They are:

(1) botanical classification,

(2) classification based on hardiness,

(3) classification based on the parts used, and

(4) classification based on essential methods of culture.

The last on is the most convenient method and is generally followed for describing the cultural operations of different vegetables. The vegetables are dealt under thirteen groups, namely potato, solanaceous fruits, cole crops, root crops, bulb crops, peas and beans, cucurbits, sweet-potato, bhindi or okra, salad crops, pot-herbs or greens, other root crops and perennial vegetables.

SOLANACEOUS FRUITS

Three important vegetables, tomato (Lycopersicon esculentum Mill.), brinjal (Solanum melongena L.) and chilli (Capsicum annuum L.) are included in this group. They all belong to the family Solanaceae.

Varieties

Tomato :

IARI—Pusa Ruby', 'Pusa Early Dwarf', 'Sioux', 'Marglobe', 'Best-of-All'. 'Slope. 120' and 'rama'

Uttar Pradesh—T 1', 'Kalyanpur Angurlata' and 'Ponderose'

Punjab—Keckruth', 'Keckruth Ageti', 'Pb. Tropical' and 'Slope 12'

Brinjal :

IARI—'Pusa Purple Long', 'Pusa Purple Round', 'Pusa Purple Cluster' and 'Pusa Kranti'

Uttar Pradesh—'T 1' to 'T 4', 'Benares Giant' and 'Black Beauty' *Bihar*—'Muktakeshi', 'S T 1' and 'S T 2'

Punjab—'Black Beauty', 'P 8' and 'P 34'

Madras—'Wynad Giant' and 'Gudiyatham'

Maharashtra—'Surti Gota' and 'Manjri Gota'

Chillies :

IARI—'California Wonder' and 'Yolo Wonder' (Vegetable type) and 'NP 46A'

Andhra Pradesh—'G 1', 'G 2' and 'G 3'

Cultivation

All the three vegetables are warm season crops. Their plants cannot withstand frost. In northern India, two sowings are done in June-July for the autumn crop, and in November for the spring-summer crop. A third sowing of brinjal can be done in March for a rainy-season crop. In the east and the south India, the crops can be grown throughout the year, the main sowing season being during July-August. In the hills, the seed is sown in March-April. For a good yield, 20-25 tonnes of farmyard manure per hectare is incorporated into the soil at the time of its preparation. This is to be supplemented with 100 kg of nitrogen, 60-80 kg each of phosphoric acid and potassium. Half the dose of nitrogen and the full dose of phosphoric acid and potassium should be added to the soil at transplanting, and the rest top-dressed after about four weeks. The foliar application of 35 kg of nitrogen and 45 kg of phosphoric per hectare as 4 to 5 sprays is also recommended. It is necessary to maintain an even moisture supply in the soil; over-watering is as harmful as insufficient irrigation. A dry period, followed by a sudden heavy watering, may cause the cracking of tomato fruits. F1 hybrids can be used for high yield and uniform fruiting. The yield of tomato is about 20-25 tonnes, of brinjal 25-30 tonnes; and of green chilli 8-10 tonnes and of dry chilli 2-2.5 tonnes per hectare.

Damping off is a common disease of all the three crops. Tomato is attacked by a number of fungal disease, e.g. Fusarium wilt, early blight and late blight. Several sprayings with fungicides, Bordeaux mixture, Marglobe and Manalucie, and the use of resistant varieties, are recommended. The common virus disease is tobacco mosaic and leaf-curl. In the case of brinjal, Phomopsis wilt is the most important fungal disease and is seed-borne. Hot-water treatment (50°C for 30 minutes) of the seed helps to reduce the incidence of the disease. Little leaf is caused by mycoplasma and can be partially controlled with antibiotics. Die-back, mosaic and leaf-curl are the most serious disease of chillies. Fruit-borer is a serious pest of tomato and so are the shoot and fruit-borer of brinjal. Thirps are important insects on chillies. These pests can be controlled with periodical sprayings with insecticides, such as Malathion, DDT or BHC.

COLE CROPS

This group includes cabbage (Brassica oleracea L. var. capitata), cauliflower (B. oleracea L. var. botrytis), knol-khol or kohlrabi (B. caulorapa L.), broccoli (B. oleracea L. var. italica), brussels sprouts (B. oleracea L. var. gemmifera Zenk.), and chinese cabbage ((B. pekinensis (Lour.) Rupr. and B. chinensis). All the above crops have developed from wild cabbage, known as colewart. They belong to the family Cruciferae and the genus Brassica. All cole crops are hardy and thrive best in cool weather, except some acclimatized early cauliflower varieties. They have many things in common in respect of their cultural requirements.

Varieties

Cauliflower :

Early-'Kunwari' and 'Early Patna'-available from mid-September to mid-October. 'Pusa Katki'—available in October-November.

Mid-season-'Aghani', 'Poosi', 'Patna Main crops'—available from mid-November to mid-December. 'early Snowball', 'D 96', 'Japanese Improved'—available from mid-December to mid-January

Late-'Dania-available in January and February. 'Snowball-16', 'Sutton's snowball'—available from mid-January to April

*Cabbage:*Round- or ball-head types—'Golden Acre', 'Pride of India', 'Copenhagen Market', 'Express Flat'. Drumhead types—'Pusa Drumhead'. Conical-head type—'Jersey Wakefield'. Savoy type—'Chieftan'

Round-head types are the earliest followed by the conical types, and then the drumhead and the savoy types.

Knol-Khol:'White Vienna', 'Purple Vienna', 'King of North'

Chinese cabbage:'Chihili', 'Wong Bok' and 'Pakchoi' (non-heading)

Broccoli:'Calabrese', 'Bronzino'

Brussels Sprouts:'Catskill', 'Long Island', 'Danish Prize'

Cultivation

The seeds are sown in nursery-beds in May and June for early, and July-August for mid-season cauliflower. The seeds of other cole crops are sown from the middle of September to the end of October. For the early crop, 600 to 750 g and for late crop 400 g seed, except for knol-khol where about 1 to 1.5 kg of seed is required, to be sown to cover one hectare. In the hills, the seeds are sown from March to June. The seedlings are transplanted when 5-6 weeks old at a spacing of 45 cm each way for the early crop. The row-to-row distance is increased to 60 cm for late varieties. In the case of knol-khol, it is reduced to 30 cm from row to row and 20 cm from plant to plant.

For best results, 15 to 20 tonnes of farmyard manure, compost or sludge should be incorporated into the soil about four weeks before transplanting. Sixty kg of nitrogen, 80 kg of phosphorus and 40 kg of potassium per hectare should be applied just before transplanting. A top-dressing of 60 kg of nitrogen, about six weeks after transplanting, at the time of earthing-up, is recommended. Water must be supplied to ensure continuous growth. An early crop may need irrigation twice a week till the start of the rains. The late crop may be irigated once a week, depending on the soil and the weather condition. Cauliflower is harvested when the curds attain a proper size and before they begin to 'rice' or discolour. Cabbage heads are harvested when they attain full size and become hard. Knol-khol knobs are harvested before they get fibrous. The average yield per hectare varies from 20 to 30 tonnes in the case of cauliflower, 30 to 40 tonnes in the case of cabbage and 20 to 25 tonnes in the case of knol-khol.

Boron deficiency causes browning or brown-rot of cauliflower. The application of 10-15 kg of Borax per hectare controls browning. Whiptail is caused by molybdenum deficiency, particularly in acid soils. Liming on acid soils or the application of 1 kg of ammonium molybdate per hectare is recommended. Damping-off of young seedlings in the nursery-bed can be prevented either by sterilizing the soil with formaldehyde or by drenching it with some fungicide. Proper drainage and aeration are necessary to prevent damping-off. Black-rot and soft-rot are caused by bacteria and can be prevented by using disease-free seeds, or with hot-water treatment of the seeds at 50°C for 25-30 minutes or by using resistant varieties. Cabbage-maggot, looper and aphids are the main insect pests and can be controlled by spraying DDT (0.1%) or Malathion (0.02%).

Seeds of early and mid-season cauliflower can be raised in the plains. The seed yield may be about 500 to 600 kg per hectare. Seeds of other crops are produced in the hills between elevations of 1,300 and 1,600 metres. Late cauliflower yields 300 kg and cabbage and knol-khol about 500 kg of seed per hectare.

ROOT CROPS

The important commercial crops grown under this group are radish (Raphanus sativus L.), turnip (Brassica rapa L.), carrot (Daucus carota L.) and beet (Beta vulgaris L.). Radish and turnip belong to the family Cruciferae, carrot to the Umbelliferae and beet to the Chenopodiaceae. The other vegetables belonging to this group are parsnip, rutabaga, salsify, charvil, skirret and celeriac. All these crops thrive well in the cool season. However, a number of varieties of radish, turnip and carrot grow well in a comparatively warm season.

Varieties

Radish, turnip and carrot varieties are divided broadly into two groups - European or temperate and Asiatic or tropical. The commonest varieties are:

Radish :

Temperate—'White Icicle, 'Pusa Himani', 'Rapid Red', 'White Tipped', 'Scarlet Globe', 'French Breakfast', etc.

Tropical—'Japenese White', 'Pusa Desi', 'Pusa Chetaki', 'Jaunpuri', 'Bombay Red', etc.

Turnip :

Temperate—'Purple Top', 'White Globe', 'Golden Ball', 'Snow Ball', 'Pusa Chandrima'

Tropical—'Pusa Kanchan', 'Pusa Sweti'

Carrot :

Temperate—'Half-long Nantes', 'Coreless', 'Chantaney'

Tropical—'Pusa Kesar'

Beet:'Crimson Globe', 'Detroit Dark Red'

Cultivation

The temperate types are sown from the middle of September to February. The tropical types are generally sown from early August to the end of October. They bolt early, if sown later. The seed-rate per hectare is 3 to 4 kg for turnip, 5 to 6 kg for beet and carrot and 10 to 12 kg for radish. Light friable soil is considered best for root crops.

The growth being rapid in a short period, fertile soil is preferred. All root crops need potassic fertilizers. To a normal soil, the application of 100 kg of nitrogen, 50 kg of phosphorus and 70 kg of potassium per hectare is recommended. All root crops require plenty of water till the roots are large enough to be pulled out. Long-rooted varieties need earthing-up at least once. Roots should be harvested when they are tender. A few days' delay in harvesting, particularly in the case of radish may make the roots pithy. The roots are pulled out along with the tops and then packed for marketing. The yield material vary from 20 to 30 tonnes per hectare.

The seeds of the temperate types are produced only in the hills. The tropical types produce seed in the plains. All the above vegetables are cross-pollinated crops. The roots left in situ produce the highest quantity of seed. However, to produce quality seed, the roots are pulled out, selected after examining their characters and then replanted.

The root crops are attacked by bacterial soft-rot, particularly in the hills. Sometimes there may be some cases of mosaic and rust. The most important insect pests are aphid and painted bug. These pests can be controlled by spraying Malathion or Folidol (0.02 to 0.03%).

Commercially important root crops such as colocasia (*Colocasia esculenta (L.) Schott.*), yam (*Dioscorea alata L.*), tapioca (*Manihot esculenta Crantz*) and elephant's-foot yam (*Amorphophallus campanulatus Blume ex Dcne.*). constitute an important source of food in the tropical areas. They are rich sources of carbohydrates. Young leaves and petioles of colocasia are cooked like other greens. The tubers of some varieties of colocasia and elephant's foot yam have acridity, which can be destroyed by boiling. However, acridity-free varieties are now available. Tapioca roots are eaten either after boiling or baking. Flour or meal made from the roots is used in making bread and dishes. Starch is also manufactured for making paper, laundering and other industrial purposes.

The root crops mentioned above are warm-season crops and are grown in summer as well as in the rainy-season. For the summer crop, they are sown in February-March and for the rainy-season crop, in June-July. A deep, rich friable, well-drained soil is ideal. Sprouted tubers of colocasia and yam (elephant's-foot) are sown in rows 45 cm apart and 30 cm within rows. Cuttings about 15 to 20 cm long are used to propagate tapioca. Most of these crops take 5 to 6 months to be ready for harvesting. Elephant's-foot takes three to four years. The yield per hectare is about 15,000 kg in the case of colocasia and 30,000 kg in the case of tapioca and elephant's-foot. Colocasia suffers from blight which can be prevented with regular spraying with Bordeaux mixture. Tapioca is often attacked by the mosaic virus. Varieties resistant to this diseases should be grown where it is a problem.

Perennial vegetables. Perennial vegetables are those which produce the edible portion continually for several years. The common vegetables under this group are asparagus (*Asparagus officinalis L.*), rhubarb (*Rheum rhabarbarum L.*), globe artichokes (*Cynara scolymus L.*) and Jerusalem artichoke (*Helianthus tuberosus L.*). These crops have tuberous roots, rhizomes or crowns. Asparagus is cultivated for its tender shoots, commonly known as spears, rhubarb for its large, thick leaf stalks, glove artichokes for its flower-bud and Jerusalem artichoke for its roots. These are all cool-season crops. The aboveground portion dies each year during winter and again in spring.

Asparagus is propagated from seeds and crowns, rhubarb from crowns, and artichoke from suckers as well as from crowns. The seedlings of asparagus are raised in well-prepared nurseries. The seedlings are transplanted when one year old. The proper preparation of soil is essential, as the crop occupies the soil for a number of years. A good amount of organic manure should be incorporated into the soil. Then 50 to 60 kg of nitrogen, 25 kg of phosphorus and 50 kg of potassium should be added annually in two doses, once in spring and again after harvesting. These vegetables start yielding sizable crops after about three years and with good care may give economic yields for about 12 years. Asparagus spears are harvested with a special knife which cuts 3 to 5 cm below the soil surface. The artichoke buds are harvested by cutting the stem 2.5 to 3 cm below the base of the bud. The stalks of rhubarb are pulled out and not cut, leaving a few stalks on the plant. New stalks come up within a week. Asparagus rust and rhubarb crown-rot and leaf-spot are some of the diseases. Asparagus beetles, garden centipedes and rhubarb beetles are the common insect pests.

BULB CROPS

The group of bulbs includes onion (*Allium cepa L.*), garlic (*A. sativum L.*), leek (*A. porrum L.*), shallot (*A. ascalonicum L.*), Welsh onion (*A. fistulosum L.*), and chive (*A. schoenoprasum L.*). All these crops belong to the family Liliaceae and genus Allium. The two most important crops commercially grown in India are onion and garlic.

Varieties

Onion :

IARI—Pusa Red', 'Ratner' and 'Early Grano'

Others—'Patna Red', 'Patna White', 'Poona Red', 'Nasik Red', 'Nasik White', 'Bellary Red',etc.

*Garlic:*There is no recognized variety of garlic in India

Cultivation

The seed of onion is sown in the nursey from the middle of October to the end of November. In the hills, the seed is sown from March to June. Eight to 10 kg of seed is sown in one hectare. The seedlings are transplanted 10 to 15 cm apart in December and early January. A crop transplanted early gives a higher yield, but the number of bolters material be high. Sometimes, small bulbs or bulbils are sown at about 1,000 to 1,200 kg per hectare for an early crop. About 20 tonnes of farmyard manure should be incorporated into the soil at the time of preparing the land. The application of 125 kg of nitrogen, 60 kg of phosphorus and 100 kg of potassium per hectare is recommended. Half the dose of nitrogen is to be top-dressed after about a month and the rest is to be applied at the time of transplanting. Weedicides, such as Tenoran at 2.5 kg per hectare in 800 liters of water, applied 3 weeks after transplanting, can control broad-leaved weeds. The crop should be irrigated so that the moisture content of the soil be kept at the optimum level. A dry spell, followed by irrigation, may cause the splitting of the outer scales. Irrigation is stopped when the tops mature and start falling. For high yields, the f1 hybrid seed is used in most of the agriculturally advanced countries. For the economic production of hybrid seed male-sterile lines are used.

Garlic is propagated by sowing cloves, the quantity used being about 350-500 kg/ha. It is sown from September to November in the plains.

Onion gives a yield of 25 to 30 tonnes per hectare and garlic about 6 to 10 tonnes. Bulbs should be thoroughly cured before storage. The kharif crop does not store well.

Onion seed is commonly produced by sowing bulbs. Bulbs 2.3 to 3 cm in diameter, weighing 1,500 kg are needed to plant a hectare and may yield about 850 kg of seed. Bulbs are planted in the beginning of October. When seedlings are transplanted in early September, most of the bulbs produce flowering stalks and form seed. The commonest disease of onion are downey mildew and purple blotch in the field, and black mould and bacterial rot in storage. The important insect pests are onion thrips and maggots.

PEAS AND BEANS

Peas (*Pisum sativum L.*) and beans occupy a position of considerable importance because of their being good vegetable and pulse crops. They are highly nutritious and contain high percentages of proteins, carbohydrates and vitamins. They all belong to the family Leguminosae. There are at least 18 types of cultivated beans. From the standpoint of green vegetables French bean (*Phaseolus vulgaris L.*), cowpea (*Vigna sinensis Savi.*), cluster-bean (*Cyamopsis tetragonoloba (L.) Taub*) and sem (*Dolichos lablab L.*) are the most important.

Varieties

Peas:

Early—'Asauji', 'Meteor', 'Early Badger', 'Arkel', 'Early December'

Mid-season and late—'Bonneville', 'New Line Perfection', 'Bridger', 'T 17', 'T 19', 'P 8', 'P 35', 'Marrowfat', 'Thomas Laxton', 'NP 29', etc. 'Early Giant', 'Lincoln' and 'Alderman' are recommended for the hills. 'Sylvia' is an edible podded variety

French bean:'Contender', 'Giant Stringless', 'Pusa Parvati', 'Tendergreen', 'Kentucky Wonder' (pole type), etc.

Cowpea:'Pusa Phalguni' (for summer), 'Pusa Barsati' (for rainy season) and 'Pusa Dofasli' (for both seasons)

Sem (Hyacinth bean):'Pusa Early Prolific'

Cluster bean:'Pusa Mousmi', 'Pusa Sadabahar', 'Pusa Naubahar', etc.

Cultivation

Peas are able to withstand relatively low temperature compared with beans. The plants, however, cannot withstand continued frost, especially during flowering and pod formation. Beans are resistant both to frost and very high temperatures. In the plains, peas are sown from the end of September to November. French bean is sown in August-September and February-March; cowpea and cluster-bean in February-march and June-July and sem in June-July. In the hills, they are sown from April to the end of May.

Seed-rate

*Peas:*early-100 to 120 kg per hectare; mid-season and late 80 to 90 kg/ha

*French bean:*Bush-85 to 95 kg; pole-25 to 30 kg/ha

*Cowpea:*12 to 15 kg/ha

These crops being legumes, only phosphatic and potassium fertilizers are generally recommended. However, a small dose of nitrogen is valuable in stimulating early growth. Farmyard manure at 20 tonnes per hectare, if incorporated into the soil at the time of preparing the land, gives good result. Twenty-five kg of nitrogen, 70 kg of phosphorus and 50 kg of potassium per hectare should be added to the soil at the time of sowing. For proper germination, a pre-sowing irrigation (rauni) is desirable. These crops are sensitive to oversupply of water. One or two irrigation at the time of flowering and fruit-setting are beneficial. Seeds of legumes, if inoculated with legume bacteria, particularly on lands not previously grown under them, give good yields.

Green pods are usually harvested at the proper stage of maturity, well-filled edible stage for peas and full-grown stage, with seeds still small, for French bean. Early varieties of peas give a yield of 3,000 to 4,000 kg and the mid-season and late varieties 6,000 to 7,000 kg per hectare. The yield of French bean varies from 4,000 to 5,000 kg for bush varieties and 7,000 to 8,000 per hectare. The yield for the pole varieties. Sem and cowpea give an average yield between 5,000 to 8,000 kg per hectare.

The important disease attacking peas and beans are Fusarium wilt, powdery mildew, rust, and pea and bean mosaic. For controlling Fusarium wilt, the best method is to use resistant varieties, and for controlling powdery mildew, dusting with sulpher is very effective. Disease-free seed should be used for controlling bean mosaic, as it is transmitted through seed. Aphids and weevils are the most important insect pests which can be controlled by spraying Malathion and DDT respectively.

CUCURBITS

This group includes cucumber (*Cucumis sativus L.*), musk-melon (*Cucumis melo L.*), water-melon (*Citrullus vulgaris Schred. Ex Eckl. & Zeyh.*), bottle-gourd (*Lagenaria siceraria*

(Mol.) Standl.), bitter-gourd (*Momordicac harantia L.*), sponge-gourd (*Luffa cylindrica (L.) Roem*), ridge-gourd (*Luffa acutangula Roxb.*), snake-gourd (*Trichosanthes anguina L.*), pointed-gourd (parwal) (*Trichosanthes dioica Roxb.*), round-gourd (tinda) (*Citrullus vulgaris var. Fistulosus Duth. & Full*), ash-gourd (*Benincasa hispida (Thunb.) Cogn.*), pumpkin (*Cucurbita moschata Dutch. Ex Poir*), summer squash (*C. pepo L.*), winter squash (*C. maxima Duch.*) and a number of other crops mostly of trailing habit. They all belong to the family Cucurbitaceae and are grown during summer.

Varieties

Cucumber:'Japanese Long Green', 'Straight Eight', 'Balam Khira', 'Khira Poona', etc.

Musk-melon:'Pusa Sharbati', 'Lucknow Safeda', 'Hara Madhum Kutana', 'Durgapur', 'Madhu', 'Arka Jeet', 'Arka Rajhans', etc.

Water-melon:'Sugar Baby', 'Asahi Yamato', 'Charleston Grey', 'Pusa Badana' (seedless), 'Tetra-2', etc.

Bottle-gourd:'Pusa Summer Prolific Long', 'Pusa Summer Prolific Round', 'Pusa Meghdut' and 'Pusa Manjari'

Bitter-gourd:'Pusa Domousmi', 'Kalianpur Baramasi', 'Coimbatore White Long', etc.

Sponge-gourd:'Pusa Chikni'

Ridge-gourd:'Pusa Nasdhar', 'Satputia'

Summer squash:'Early Yellow Prolific', 'Australian Green', 'Butternut', etc.

Winter squash:'Arkha Suryamukhi'

Tinda:'Arka Tinda'

Cultivation

The culturalreqs of all the commercially important crops in this group are more or less similar. Cucumber, bottle-gourd, bitter-gourd, pumpkin, sponge-gourd and ridge-gourd can be grown in summer as well as in the rainy season, whereas, musk-melon, water-melon, squashes and tinda grow better only in summer. Winter squash grows well under mild climatic conditions.

There are two methods of sowing. Ridges are prepared at proper spacing and after adding manure, a number of seeds are sown on each ridge. In the other method, furrows are made and seeds are sown on the edge either on one or both sides. The spacing from row to row, unless staked, varies from 1-1/2 to 3 m, according to the crop. The distance from plant to plant is kept at 60-90 cm. Approximately 2.5 to 3 kg of seed is required for sowing a hectare of cucumber, musk-melons and water-melon, 4 to 5 kg of seed of bottle-gourd, sponge-gourd and ridge-gourd, and 7 to 8 kg of seed of pumpkin and squashes.

The summer crop is sown from January to March and the rainy-season crop in June-July. In the hills, they are sown in April. About 30 tonnes of farmyard manure are added to the soil at the time of preparing the land. For a good yield, 80 kg of nitrogen in two doses and 50 kg each of phosphorus and potassium per hectare should be applied. The average yields per hectare are:bitter-gourd and squashes, 6,000 to 8,000 kg; cucumber and musk-melon, 8,000 to 10,000 kg; water-melon and bottle-gourd, 15,000 to 20,000 kg, and pumpkin, 20,000 to 25,000 kg.

The cucurbits are also grown on the river-beds during summer. Special techniques are followed to get an early crop on the sandy banks of the rivers.

Cucurbitaceous crop are attacked by a number of diseases, of which powdery mildew, downey mildew, Fusarium wilt and virus diseases are of economic importance. Powdery and downey mildews can be controlled by spraying the crop with Karathane or Morestan, or by growing resistant varieties. Red pumpkin beetle and fruit fly are the most important insect pests. Dusting with 1% Lindane or a mixture of BHC and DDT control the beetle, but DDT should be dusted with caution, as it may damage the young bit leaves.

SWEET POTATO

The sweet potato (*Ipomoea batatas (L.) Poir.*) is a very important crop in the tropical regions of India. The chief uses of sweet-potato are for human consumption and for the manufacture of starch and alcohol. It contains about 16 per cent starch and about 4 per cent sugar. It belongs to the family Convolvulaceae.

Varieties

'Pusa Suffaid', 'Pusa Lal', 'Pusa Sunheri', 'SP-3', 'SP-9', 'Ranger', 'Bhadrakali', 'Hosur Red', 'Gold Rush', 'Centennial', etc.

Cultivation

Sweet-potato requires a long and warm growing season. It does not stand frost. A moderate proportion of sand in the top soil, with a fairly retentive subsoil, provides ideal condition for its growth. It is grown from sprouts produced from its tuber-like roots and from vine cuttings. In well-prepared nursery-beds, the selected roots are planted 30 cm apart in rows which are spaced 45 cm. The sprouts are cut and planted for further growth in a second nursery. Ultimately, the cuttings from this nursery are planted at about 60 cm from row to row and 30 cm within the row. About 40 to 50 thousand cuttings are required to plant one hectare.

In northern India, the cuttings are planted during June-July and in central and southern India during October-November. In some parts, both kharif and rabi crops are grown. A mixture of about 60 kg each of nitrogen and phosphorus and 120 kg of potash per hectare may be applied to obtain good yield. Both flat beds and ridges are used in various parts from 10 to 15 thousand kg per hectare. Red-skinned roots generally store better than white-skinned ones.

The commonest diseases is stem-rot which is spread through soil or diseased roots or vines. The best method of conteolling the diseases is to use diseases-free planting material, and resistant varieties and follow a long crop rotation. Sweet-potato weevil and leaf-eating caterpillars are the most damaging insect pests. The control measures consist in planting roots and cuttings free from weevils and dipping the cuttings before planting into a DDT solution or a mixture of 500 g of lead arsenate in 50 liters of water and spraying the crop with lead arsenate at fortnightly intervals.

OKRA OR BHINDI

The okra(*Abelmoschus esculentus (L.) Moench*), belongs to the family Malvaceae. It is cultivated throughout India for its immature fruits. Its varieties are:'Pusa Sawani', 'Pusa Makhmali', 'Perkin's Long Green' (for hills only), etc.

Cultivation

Bhindi plant prefers a long warm season. The plant is tender and cannot tolerate cold at any stage of growth. Low temperature true in the early spring results in poor germination of seed. Bhindi seed does not germinate at 20°C or below. The best germination takes place between 24°C nad 30°C.

Soil and Soil Preparation

Bhindi can be grown on all kinds of soil except light sandy soil. Well-drained loamy soil rich in organic matter is, however, preferred. Presence of sufficient amount of organic matter improves the nutrient status, soil structure and water holding capacity. Operations for the preparations of land for bhindi sowing would depend upon the condition of the plot. If the plot was under crop which left behind stubbles and the organic material, it has to be ploughed or disced before cultivator is used. In such land normally the disc should be used twice and the tiller three or four times. Two to three plankings would give the desirable structure of the soil. However, if the land was under a crop as potato, one disc operation followed by 3-4 times cultivation would produce the desirable condition of the soil. However, the land has to be planked at least twice before the final sowing is taken up. Sowing is done by two methods :

1. Sowing on ridge
2. Flat sowing

Sowing on Ridges

This method is particularly important for the early crop sown in february. the field is divided into plots and within each ridges are made 45 cm apart. If possible these should be made running East-West. Seeds are dibbled 1 cm below the soil surface keeping a distance of 15 cm between hills. seed is dibbled on the top of the ridge or slightly on the side facing the sun. This is very important for early crop, as the side facing the sun attains a temperature several degrees haigher than that on the other side. This high temperature affects the seed germination favourably. Ridge sowing for early crop has another advantage. Seeds take many days to germinate and often an irrigation is required to keep up the right moisture content in the soil. In this case pre-emergence irrigation does not lead to crust formation as water is not allowed to run over the ridges.

Flat Sowing

This method is used for the later crop when the days are quite warm. Sowing is done in rows 45 cm apart, which have been previously drilled with fertilizers. Single row cotton drill can also be used efficiently for flat sowing. This method is quite popular with bhindi seed growers. The advantage of this method over ridge sowing is that it is less labour consuming. Bhindi seed has a hard seed coat. In order to get good germination it is always advisable to soak the seed overnight before sowing. For early crop the soaked seeds may be kept covered at some warm place for sometime so that germination initiates there and slightly sprouted seeds can then be planted in the field.

Sowing Time

Bhindi is soen twice a year in the plains for green pods. For early crop the seed is sown from February to April and for late crop in June-July. For seed production the second crop is most suited. It is not only gives higher yield but seed is also of good quality and also provides an opportunity for roguing of the virus susceptible plants. During dry weather virus does not develop and the undesirable plants are not detectible. Period commencing from last week of June to first week of July has been found to be the best. In case the seed crop is sown in May or early June then the main fruiting period coincides with mansoon rains. Water gets entry into the ripe dehisced fruits and spoils the seed within the pods.

Seed Rate

Temperature has pronounced effect on germination of bhindi and consequently the quantity of seed used to obtain a reasonable crop stand depends on the prevailing temperature. To start with early in the season the seed requirement is as high as 15-18 kg per acre and it goes down to 8-10 kg in March. For the second sowing i.e. june-July 5-6 kg seed is sufficiently for one acre. Seeds older than two years lose their germination and should be avoided.

Manures and Fertilizers

Fifteen to twenty tons of well rotton farmyard manure or compost should be applied in an acre of land. The quantity should be increased on poor and light soils. Addition of farmyard manure is particularly important for the early sown crop. Farmyard manure raises the soil temperature which helps the seed germination. Farmyard manure should be applied while preparing the seedbed so that it gets well mixed up in the soil.

Farmyard manure should also be supplemented with chemical fertilizers. The quantity of the fertilizer required should preferably be ascertained from the condition of the crop rather than strictlyadhering the recommendations. Pale yellow colour of leaves and stuned growth is an indication of the nitrogen deficiency in the soil. On an average 144 kg per acre of calcium ammonium nitrate or 80 kg of urea should be applied. In seasons of excessive rainfall an addition does of nitrogen may be benificial. Deficiency of nitrogen results in poor plant growth and consequently low fruit yield, small sized fruits, and loss of tenderness of the fruit.

Bhindi usually does not respond to the application of potash. However, on soils where the nutrient status is low, potassium sulphate at the rate of 50 kg per acre may be applied before sowing. application of potash to seed-crop results in plump seeds.

Irrigation

First irrigation in case of ridge sowing should be given immediately after sowing. Care should be taken not to allow the ater to overflow the ridges. In case of flat sowing first irrigation should preferably be given only after the seedings have come up. Subsequently irrigations should be given after every four to five days in the hot season or every 10-14 days in moderate season. On loose sandy type of soils frequency of irrigation should be increased.

Hoeing and Weeding

In order to keep the weeds under control and to maintain desirable soil structure, three to four hoeings should be given. First hoeing may be given when the seedings are two week old and subsequent hoeings may be repeated at fortnightly intervals. Hand hoe can be used efficiently and economically for this purpose. Second crop, i.e. sown should be earth up to avoid the damage from water stagnation. Weeds can also be kept under check by the use of herbicide Basalin, sprayed @ 800-1000 ml per acre as pre-plant application four days before sowing. The herbicide should be incorporated in the soil by harrowing, Another weedicide Lasso can also be used @ 2 litre per acre as pre-emergence spray one day after sowing or stomp one litre per acre or 750 ml acre+one hoeing as pre-emergence spray are recommended. One hoeing may be necessary after 60 days if the intensity of weeds is high.

Harvesting and Marketing

The fruits attain marketable size when the plants are 45-50 days old, depending upon the temperature. It takes ficve to seven days to form an edible fruit after the opening of the flower. The fruits should not be allowed to over grow to give a fibrous appearance because overgrown fruits are not liked by the consumer. Only tender fruits should picked for better returns. Leaving fruits on the plants for a longer period not only impairs the quality of the fruit but also reduces the rate of apical growth and thus ultimately reflects upon the fruit bearing capacity of the plant. Therefore, frequent picking of fruits is desirable. Picking should be done early in the morning because fruits make a lot of growth during night. If at all fruits are to be picked in the evening it should be in the late hours.

Water should be sprinkled on the harvested fruits to keep them fresh. Harvested fruits kept at room temperature deteriorate in quality as vitamin C is lost rapidly. Fruits should be graded keeping in view the size, shape and colour and packed in basket or wooden crates for disposal in the market. Fruits should not be carried in gunny bags as this will lead to brusising of fruits which spoils their look. small, tender, green fresh and unbruised fruits fetch a premium in market. A yield of 4500-5500 kg of green edible fruits can be obtained from an acre of land. Okra fruits are not suitable for cold storage for a long time. However, the fruits can be kept in storage at 50° to 10°C and relative humidity 85 to 90 percent for about 10 days.

Varieties

There are four varieties which are recommended for cultivation.

Punjab-8 :

The plants of punjab-8 are medium tall with splashes of purple pigmentation present on the stem. Leaves are deeply lobed and less serrated. Leaves, stem and petiolesare less hairy. Fruits are thin, long dark green and five ridged. It has got a high degree of resistance to yellow-veins-mosaic virus and tolerance to jassid and borer. It is suitable for processing. Average market yield is 55 q/acre.

Punjab-7 :

This is the latest variety of bhindi developed at Punjab Agriculture University, Ludhiana possessing better resistance to yellow vien mosaic virus thanthen any other variety in the country. Plants are medium tall and green. Leaves are deeply lobed and hairy. The purple hue may or may not be evident on stem and petiol but characteristically evident at eh base of leaf lamina. Fruits are long, five ridge and green with a pale green colour at the fruit base. It takes 50-55 days from sowing to fruiting. Average yield is about 45 q per acre.

Punjab Padmini

Plants are tall and green with mils purple tinge on the stems and leaf petioles. Leaves are dark green, deeply lobed and hairy. fruits are quick growing, dark green smooth, thin, long, five ridged and retentive of tenderness. Fruiting starts after 55-60 days after sowing in the spring crop. It has field resistance to yellow vein mosaic virus. Average yield is 45 q per acre.

Pusa Sawani

This is an old variety of bhindi and has a wide popularity all over the country. It has been developed by I.A.R.I., New delhi, and used to show field resistance to yellow vein mosaic. Previously it was recommended as a rainy season variety but now it is recommended for spring crop. Fruits are five ribbed, dark-green in colour and free from bristles.

Pests

Spotted Boll Worm:The larvae of this moth attack the shoots of plant by boring into it. It also bores into the fruits which are rendered unfit for human consumption.

Jassid:The adults and nymphs of this insect feed on the palnts by sucking the sap. Leaves turn pale and curl upwards. The cupping may be followed by drying of leaves from the margins giving a characteristic sympton known as hopper burn. Both spotted boll-worm and jassid attack the crop from May to September.

For controlling both these pests spray at fortnightly intervals with 500 ml Malathion 50 EC in 100-125 litres of water per acre. As soon as fruiting starts, spray 500 g of carbrayl 50 W.P. or 350 ml Thiodan 35 EC or 100 ml of Sumicidin 20 EC (fenvalerate) or 80 ml of Cymbush 25 EC (Cypermethrin) in 100-125 L of water per acre. Picking of fruits should be done before spraying. Borer infested fruits if any, should be removed regularly and buried deep.

Spider Mite : The plants are considerably damaged and weakened tha adults and nymphs suck the plant sap. In severe cases the leaves dry up and fall off.

The crop should be sprayed with 250 ml of Metasystox 25 EC or Rogor 30 EC in 125 litres of water.

Diseases

*Wilt:*The growth of the plant becomes stunted and the leaves give the appearance of yellow and wilted growth. The stem turns dark, close to the soil surface and finally the entire plant wilts.

Bhindi should not be rotated with crops like tomato, eggplant and chillies. It should not be grown in the infested field for three years. Use seed from healthy plants only. Treat the seed with 3 g of Captain or Thiram per kg seed before sowing.

*Yellow vien Mosaic:*During rainy season, it is a very serious disease. Due to clearing of veins, leaves look chlorotic. There is a yellow and mosaic pattern. Fruit production is adversely affected. Fruits become yellow and lose acceptibility.

Bhindi varieties with narrow leaves should be planted instead of varieties with broad leaves. Disease resistant varieties like Panjab-7 and Panjab Padmini should be sown. The insects such as white flies, responsible for the spread of this disease, should be controlled. Spray 560 ml of Malathion 50 EC in 350 litre of water per acre for the control of vector.

SALAD CROPS

Some vegetables are consumed in the uncooked state and are known as salad crops. They are lettuce (Lactuca sativa L.), celery (Apium graveolens L.) and parsley (Petroselinum crispum (Mill.) Nym). All of them are cool-season crops and are mostly grown around big cities. The leaves of lettuce and leaf-stalks of celery are eaten, whereas parsley leaves are used for decoration, garnishing or in soups, sauces and stews.

Varieties

Lettuce:'Great Lakes', 'Chinese Yellow', 'Slobolt'

Celery:'Satndard Bearer', 'Wright Grove Giant'

Parsley:'Mosscurled', 'Hamburg'

Cultivation

The seeds are generally sown in the nursery-beds from August to October. About 500 g of lettuce, 250 g of celery and 1.5 kg of parsley seeds are required to raise seedlings for one hectare. The seedlings of celery are often transplanted in well-manured trenches. Blanching is often done by wrapping-paper or black polythene around the leaf stalks or by earthing up the soil as the plants grow. The leaves of lettuce and parsley can be used even when they are young. Each celery plants is cut just below the surface with a sharp knife. The diseases which affect lettuce are slimy soft-rot and mosaic and those affecting celery are ear;y and late blights, pink-rot, mosaic and aster yellows. Aphids attack both lettuce and celery. It is safe to use 3% Nicotine dust on the leaves. Celery leaf-miner and carrot weevil are other pests.

Leafy vegetables, potherbs or greens

Leafy vegetables are very rich in minerals and vitamins A and c. They also supply the roughage required in our daily diet. A number of greens are cultivated in India. The leaves of some shrubs and trees are also used as greens. These vegetables are grown throughout the year. Some are suitable for growing during winter e.g. palak (*Beta vulgaris L.*), spinach (*Spinacia oleracea L.*), fenugreek, methi (*Trigonella foenumgraecum L.*) and mustard sarson, Brassica spp). Others, such as amaranthus (*Amaranthus spp.*), portulaca (*Portulaca oleracea L.*) and poi (*Basella alba L.*) are suitable for growing during summer. Other leafy vegetables, which are also eaten in different parts of the country, are karam saag, New Zealans spinach, buckwheat, bathua, chakwat, etc.

Cultivation

The land is laid out in plots convenient for irrigation. A basal dressing of 35 to 40 tonnes of farmyard manure should be incorporated into the soil at the time of preparing the land. A mixture of 40 kg of nitrogen, 30 kg of phosphorus and 30 kg of potassium per hectare is applied to the soil at the time of sowing. The seed is sown either broadcast or in rows about 15 to 30 cm apart and the surface is raked to cover the seed. Twenty-five to 30 kg of seed of palak, spinach and fenugreek, 6 to 8 kg of mustard and about 2.5 kg of amaranth are required to sow one hectare. One light irrigation may be given immediately after sowing, followed by subsequent irrigations at 8- to 10-day intervals. The average yield may vary from 7,000 to 10,000 kg of green leaves per hectare. The important diseases which attack leafy vegetables are damping-off, Cercospora leaf-spot, mildew and rust. These diseases can be controlled by treating the seeds with Ceresan and by spraying some copper fungicides, e.g. Bordeaux mixtures. The common insects are aphids, caterpillars and beetles.

REFERENCES

Bailey, L.H., *Principles of Vegetable Cultivation*, Reprint. 1999.

Jacob John, P., *A Handbook on Post Harvest Management of Fruits and Vegetables*, Delhi, Daya Publishing House, 2008

Rai N. and Yadav, D.S., *Advances in Vegetable Production*, New Delhi, Researchco Book Centre, 2005.

Romesh Kumar Sud and Sudhir Kumar, *Flowers and Vegetables of India*, Jodhpur, Scientific, 2004

6

Production of Ornamental Crops

Flowers are symbolic of beauty, love and tranquility. Besides their aesthetic value, they are important for their economic uses, such as for cut-blooms and for extracting perfumes and other products. In our country, flowers are sanctified and are commonly used in worshipping the deities in our homes and temples. We are intimately associated with them, and on all festive occasions like marriages, religious ceremonies and social functions, the use of flowers and garlands has become almost essential.

According to an early survey made by the Indian Council of Agricultural Research, flowers are grown in about 4,000 hectares for commercial purposes, with an annual production of about 10,500 tonnes of cut-flowers worth 9.26 crores of rupees sold annually in the markets of Bombay, Calcutta, Madras , Bangalore and Delhi. However, with the increase in demand for cut-flowers in recent years, the production is estimated to have gone up considerably. Consequent upon the rapid development of hotels and tourism, there has been a sudden rise in the number of florist's shops and kiosks in the metropolitan cities. The perfumes extracted from some flowers, e.g. rose and jasmine, are no longer luxuries, as considered by many earlier, but they have become the essential needs of our daily life. Perfumes are used in manufacturing soaps, cosmetics, hair-oil and in food and tobacco industries.

Besides, the seed and nursery business is a source of income to many for their livelihood and provides employmnent for many. There is also a considerable scope for exploiting the export potential of ornamentals, particularly cut-flowers, plants,seeds and bulbs. Recently, there have been some sporadic, but unsteady, attempts also to export ornamentals.

There are some other perennial vegetables, such as cho-cho (*Sechium edule L.*), sea-kale (*Crambe maritima L.*) and horse-radish (*Armoracia rusticana*) which are grown in a very limited area. There are also some fruit crops, e.g. banana, papaya and jackfruit, which are cooked as vegetables in the immature stage. The drumstick is also a popular perennial vegetable.

Botanically, the ornamentals are spread over the entire plants kingdom, including both lower and higher plants, herbaceous annuals, biennials or perennials or woody trees, shrubs and climbers, succulents, desert or aquatic plants and epiphytes and terrestrials. Of the large variety of ornamentals grown in the gardens of India, a few very important flowering plants, useful for garden display, cut-flowers or the perfume industry, are described in this chapter.

ROSE

The modern roses are derived from crosses between the Chinese roses (*Rosa chinensis Jacq.*) and the European roses (*R. gigantea, R. damascena Mill., R. moschata Herrm*). There are six different types of modern roses, grown in gardens, namely, hybrid-tea, floribunda, polyantha, climbing and rambling, miniature and shrub roses. The flowers in the hybrid-tea are well-shaped and are borne singly, whereas in the floribunda they are large and are produced in clusters, unlike those of polyantha, which has smaller flowers in larger trusses. The miniatures are dwarf, having small leaves and flowers, and are suitable for growing in pots.

Propagation

Roses are propagated both by budding and by using cuttings but the former method is preferred for obtaining better-quality and faster-growing plants. In Bengal, roses are also propagated by inarching. The rambler, polyantha and miniature roses are successfully multiplied from cuttings. The cuttings are inserted into the soil or sand from December to February. The commonly used rootstock is the Edouard rose, whereas in Bengal, in the northern hills and in southern India R. multiflora is used for this purpose. The rooted plants of the rootstock are transplanted in July or August at sites where the budding is to be attempted. The best time to bud roses is from December to February in northern India.

Planting

The rose-bed should be located in a sunny situation, away from trees or hedges and it must receive sunshine at least during the whole forenoon, if not during the whole day. The bed should be well-drained, as the rose does not thrive in a wet or water-logged soil. The best time to plant roses is from September to November in the northern plains of the country. The bushes are planted about 60 cm apart in a row and the distance between the rows is 76 cm. The standard or tree-roses (budded) at a height of 1 metre are planted about 1-1.25 m apart, whereas the distance between the climbers may be about 2-2.5 m. It is useful to apply about 8-10 kg of cowdung manure and 3 g of Aldrin or BHC 5% to each pit at the time of planting.

Pruning

The rose plants are pruned once a year during the second or third week of October in the northern plains. After about 6 to 7 weeks of pruning, the plants start flowering. The time of blooming can be adjusted according to the date of pruning. The new or so-called "maiden" plants are not pruned and these are generally tipped lightly before planting. In the old hybrid-tea bushes, the previous season's thick shoots are pruned up to half their length, keeping about 5 to 6 eyes on each stem. A slanting cut is made a little above an eye which is facing outwards. The floribunda is pruned moderately. The climbing and rambling roses do not require any pruning, except the removal of unhealthy, dead and interlaced twigs. The polyantha roses are pruned lightly, whereas the miniatures are generally not pruned.

Manuring

After pruning, the soil in the bed is dug up with a fork with due care to avoid any damage to the roots. About 8-10 kg of cowdung manure is applied to each bush by incorporating it into the soil, and then the plants are watered copiously. After a fortnight, a mixture of fertilizers may be top-dressed. A convenient and useful fertilizer mixture can be prepared, by taking the weight, ammonium sulphate (2 parts) or urea (1 part), superphosphate (8 parts) and potassium sulphate (3 parts), and about 100 g of this mixture may be applied to each bush.

A top-dressing of the fertilizer mixture can be given again in January-February after the first flush of flowering is over. A foliar spray of 2 parts of urea, 1 part of dihydrogen ammonium phosphate, 1 part of potassium nitrate and 1 part of potassium phosphate is also useful. About 15 g of this mixture may be added to 2.5 liters of water for spraying. It is often useful to spray the foliar feed, along with an insecticide, e.g. Malathion or Masurdin. The foliar spray may be started by the middle of November and continued till the flowers open but it should not be applied when the plant is in full bloom, as it will damage the flowers.

Watering

Heavy watering at comparitively long intervals is more useful than frequent light watering. Water-logging is harmful to roses.

Suckers

The suckers or shoots of the rootstock emerging from the base of the plant should be removed as soon as they appear. They can be distinguished from those of the scion by the shape and the size of their leaves.

Diseases and Insect Pests

The common diseases of roses are black powdery mildew and black spot. Die-back results in the blackening and drying of shoots from the cut-ends downwards after pruning. One spraying with rogor (0.1%) soon after pruning, followed after a week by another spraying with Captan (0.2%) is effective in controlling this diseases. Benlate (0.1%) may be sprayed from February to April to control powdery mildew and again in November to control black spot.

Among the insect pests, the important ones are aphids, thrips, chafer, beetle, red scale, mites and termites. The aphids, which appear during December to March, can be controlled by spraying the plant with Malathion (10cc in 10 liters of water). In July-August, the plant may be sprayed with DDT (0.2%) to control the chafer beetle and thrips. From August to October, and again in April, a spraying with Parathion (0.1 per cent) is effective in controlling the red scale. Aldrin-dust (5%) at the rate of 30 g per m2 is useful against termites.

Varieties

There are several thousand varieties of roses and several hundred new ones are being added every year. The choice of varieties depends mainly on the climate and the soil of the growing region for cut-flowers, exhibition, garden display, etc. and on personal or family preferences. A few important varieties are mentioned below.

— *Red and dark red*:'Avon', 'Papa Meilland', 'Oklahoma', 'Mister Lincoln', 'Christian Dior', 'Happiness'
— *Orange*:'Hawaii', "Super Star'
— *Yellow*:'Summer Sunshine', 'King's Ransom', 'Kiss of Fire', 'Golden Splendour', 'Golden Giant'
— *Pink*:'Eiffel Tower', 'Michelle Meilland', 'Peter Frankenfeldt', 'First Prize', 'Montezuma' (coral pink), 'South Seas'
— *White*:'Virgo', 'Matterhorn', 'John F. Kennedy', 'Dr. Homi Bhabha'
— *Bicolours*:'Bajazzo', 'Ingo Hortsmann', 'Rose Gaujard', 'Granada'
— *Lavender or Mauve*:'Blue Moon', 'Lady X'
— *Copper colour*:'Thais', 'Vienna Charm', 'Whishky Mac'
— *Striped*:'Anvil Sparks' (red with yellow streaks), 'Careless Love' (pink with white streaks)
— *Fragrant*:'Crimson Glory' (red), 'Papa Meilland', 'Oklahoma', 'Fragrant cloud', 'Charles Mallerin' (Velvety red)
— *Floribunda*:'Charleston' (yellow and crimson), 'Flame-neo' (salmon), 'Orange Sensation' (deep orange), 'Banjaran' (red and gold), 'Iceberg' (white), 'Arthur Bell' (yellow), 'Africa Star' .(mauve), 'Else Poulsen' (pimk, single), 'Delhi Princess' (pink), 'Himangini' (white), 'Prema' (soft-pink with deep-pink edges), 'Summer Snow' (white), 'Dearest' (rosy salmon), 'Zorina' (gealdine red) and 'Zambra' (orange)
— *H.T. Type Floribunda*:'Queen Elizabeth' (pink), 'Sea pearl' (pink), 'Pink Parfait' (light pink), 'Tiki' (shell pink)

— *Polyantha:*'Chatillon Rose' (deep pink, single), 'Vater Tag' (vermillion), 'Echo' (pink changing to white)
— *Miniature:*'Cri Cri' (salmon coral), 'Baby Masquerade' (lemon chrome), 'Little Buckaroo' (velvety red), 'Rosemarin' (silver rose), 'Coralin' (red orange)
— *Climbimg:*'Clg. Show Girl' (pink), 'Clg. Virgo' (white), 'Clg. Summer Snow' (white), 'Prosperity' (white), 'Marechal Niel' (Lemon), 'Golden Showers' (yellow), 'Lamarque' (white), 'Mardan White' (white), 'Mardan Pink' (pink), 'Cocktail' (red with yellow centre), 'Josephs Caat' (red and yellow)

Rose Perfume

The oil of rose is extracted from Rosa damascena Mill variety. triginipetala Dieck (2n=14), popularly called Kazanlik Rose or 30 petalled rose. Bulgaria produces the largest quantity (2 tonnes) and the best quality of rose-oil in the world. The current price of oil is about 3,800 dollars per litre (approx. Rs 32 per g).

In India, the Damask rose, Rosa damascena, or the Bushra rose similar to the kazanlik rose of Bulgaria, is grown in several areas in Uttar Pradesh, namely Barwana, Banwaripur, Ban, the villages of Hasayan and Badanpur in the Aligarh district, Sikandarpur in the Ballia district, Jhinwar (Etah district), Kanauj, Ghaziapur, Kanpur, Saharanpur and Jaunpur districts and also in Kannaur in Haldi Ghati near Udaipur (Rajasthan). The largest areas (400 ha) is in the aligarh district, where the annual production is estimated at about 7,400 quintals. It is grown as an irrigated crops. Flowering occurs during March-April for about 20-25 days once a year. The bushes are pruned in December and at this time, the cuttings are planted in the field for raising a new crops. Very little fertilizer is used, except some farmyard or cowdung manure at the time of pruning. The yield of flowers is up to 3,600 kg per hectare, depending upon the age of the bushes.

The oil percentage in a rose flower is about 0.045. The flowers are plucked in the morning for the extraction of oil. The distillers, who usually come from Kanauj, set up the distillation apparatus in the cultivators fileds in the Aligarh district, where the rose flowers are brought from the neighbouring areas for extracting oil. The common products prepared are rose-oil, gulkand and rose-water. Rose concrete is produced in very small quantities. The Indian rose-oil compares favourably with the Bulgarian ose-oil. The quality of rose-oil depends upon the percentage composition of alcohols, viz. citronellol fraction is always more than that of geraniol in the ratio of 1.2:1 to 1.5:1. The Kazanlik rose-oil contains 72 and 74 per cent of these alcohols.

Another variety grown is R. damascena variety bifera. It is commonly cultivated in Kanauj (Uttar Pradesh) and gives a second flush of flowering in September. However, its oil is inferior in odour. The Edouard Rose (R. borboniana) or Chiniya gulab is grown for making gulkand , but it fetches a lower price. The species R. centipetala is also grown in some parts of India for making gulkand and rose-water. Its oil is also inferior both in content and quality.

JASMINE

About 200 species of Jasminum, both climbing and trailing or shrub, with erect habit are found in temperate, tropical and subtropical regions of the world. Several species of Jasminum, including many important cultivated ones, are native of India.

The jasmines are highly prized for their fragrant flowers, used in the preparation of perfume and concrete. The jasmine flowers are commonly used for making garlands and 'veni' for adorning the hair of women. In the garden, the jasmine is raised as a dwarf-growing shrub, and a few species are grown as climbers.

Species

The commonly grown species are Jasminum officinale, J. grandiflorum (chameli or janti), J. sambac (bela, motia, mogra, donthara malle, boddu malle, moturia, rai or madanban), J. auriculatum, J. augustifolium, J. paniculatum, J. pubescens, J. arborescens and the yellow-flowered J. primulinum and J. humile. The flowers of J. grandiflorum are used for extracting the essential oil. J. sambac is most commonly grown in the gardens. Its flowers are single, semi-double or double, and large or small, white and fragrant. Almost all species flower during summer and the rainy season. The species J. pubescens flowers during winter in northern India.

Propagation

The jasmines are commonly propagated by layering or by using the cuttings of almost mature wood.

Cultivation

Planting is done in the rainy season. The dwarf shrub-like J. sambac is planted about 1-1.5 m apart whereas the creeper may be grown about 3.5 m apart, trained on arbours, arches, pegolas, screens or walls. J. sambac can also be grown in large pots. The plants may be pruned after flowering to obtain better growth and better flowers next season.

In Tamil Nadu, Andhra Pradesh an Karnataka, the jasmines, particularly J. sambac, are commercially cultivated for cut flowers. The watering of the plants is withheld in the end of November to bring them rest and shed their leaves. Sometimes, the foliage is also removed by hand. The shoots are also pruned to half their length in January, and cowdung or farmyard manure, at the rate of about 10 kg per bush, is applied after exposing the roots for a few days. The watering of the plants is started slowly and later increased after the appearance flower-buds. Watering is withheld after each flush of flowering until fresh flower-buds appear again. The flowerin gin J. sambac is best during summer, particularly in June-July. The creepers bloom for a longer period or almost throughout the year. Its flowering takes place in flushes at intervals of one week. The yield of flowers is about 1,000 to 1,500 kg per acre. The climbing types yield a little more, about 2,000 kg per acre.

CHRYSANTHEMUM

In popularity, chrysanthemums are perhaps next only to roses and have been in cultivation for more than 2,500 years. They have undergone remarkable changes as the result of artificial crossing and selection in their native countries, namely China, Japan, England, France, the USA and Australia. There are thousands of varieties now in cultivation in different countries. It is a national flower of Japan, where its cultivation has reached its perfection. It is very commonly grown in glass-houses for cut flowers and its flowering can be manipulated with artificial light and with controlled temperature.

Classification

Chrysanthemum varieties are classified into seven main groups, namely, incurved (a perfect ball), incurving (petals incurve loosely or irregularly), reflexed (dropping florets), Anemone (single petals with a tubular central disc), Pompon (very small flowers, without any visible centre), singles (five petals with a central disc), miscellaneous, such as Spider (petals with a hook at the tip), Spoon (with a spoon-like tip of petals), Koreans (small single, semi-double or double flowers with a visible disc) and Rayonnantes (with quilled petals).

In India, Chrysanthemum indicum, small-flowered (yellow or white), is used for commercial cultivation for cut-flowers in southern India, particularly near Coimbatore, Madurai and Bangalore.

Cultivation

The plants may be grown, both in pots and in the ground. During early February, the plants are beheaded after the flowering is over by cutting the stem, about 15-25 cm above the soil. After some time, when new suckers appear at the base of the stool, they are separated and planted in small 10 cm pots. Each sucker has its own roots. The potting-mixture consists of one part each of sand, soil and leaf-mould and a trace of wood ashes.

The second potting is done at the end of April and the suckers are transferred to bigger pots (15 cm) which are filled with a richer soil mixture containing one part of sand, one part of soil, two parts of leaf-mould, a quarter part of wood ashes and one tablespoonful of superphosphate. The third and final potting is done in August when the plants are shifted to 25-30 cm pots.

The potting-mixture consists of one part of sand, one part of soil, two parts of leaf-mould, two parts of cowdung manure, a quarter part each of small pieces of wood charcoal and wood ashes and two tablespoonfuls of superphosphate. During May-June, the young plants should be protected from strong sunshine and from heavy rain during the rainy season.

Sometimes, young cuttings can be taken in July-August from the lateral growths in the leaf axils for propagation. The cuttings are about 5-8 cm long and their cut-ends are dipped into a root-promoting hormone, e.g. Seradix B, before planting them in sand. The lower leaves of the cuttings are removed before planting.

Stopping and Disbudding

By late May or early June, the young plants are pinched (stopped) by removing the tip of the main stem at a time when the young lateral shoots or 'breaks' are just appearing in the leaf axils. Stopping induces lateral growths to develop from the leaf axils, and the number of main stems one would like to keep can now be decided upon. Generally one, three or six stems are retained for obtaining exhibition blooms. At the end of each stem, there develops the first crown bud which is allowed to develop and the lateral growths, arising from the leaf axils are disbudded or removed.

The terminal end of a stem usually has three flower-buds, the central one, being large, is the crown bud, with two small buds on each side. It is the crown bud which is allowed to develop, whereas the others are removed. However, in the case of Pompons, Singles, Koreans and Sprays, no disbudding is practiced. Sometimes, in the case of a variety the first crown buds are removed and the second crown buds are secured to obtain flowers.

These second crown buds, in general, produce smaller flowers, but of a more intense colour, particularly so in the case of oink varieties. The date of blooming depends mainly upon the time of starting the suckers or cuttings and the dates of stopping and disbudding. In foreign countries, e.g. Japan, the USA and the UK, the exact dates of stopping and disbudding have been ascertained for each variety to produce best quality flowers. The plants need staking during October. In Japan, the plants are trained in different styles.

Manuring

Ammonium sulphate (30-35 g) or urea (15g) may be mixed with 9 liters (2 gallons) of water and 0.5 litre (pint) of the mixture may be applied to each plant during July-August. Soon after the appearance of flower-buds, sulphate of potash may be applied in the same way as ammonium sulphate. About one tablespoonful of superphosphate may be mixed with the soil at the time of final potting. The liquid manure can be applied once a week after the appearance of flower-buds till the flowers are half open. Over-feeding is harmful, and it can be judged by snapping a leaf into two. If the leaf is dark green and brittle and breaks into two clean halves, further feeding of plants should be stopped.

Watering

The young plants require frequent watering in summer but less during the rains. Over-watering should be avoided. The pot which sounds heavy on tapping with a wooden hammer indicates that it needs less watering, whereas the one with a clear sound will require watering.

Pests and Diseases

The grubs of chafer beetle appear usually in July-August at the base of the pot and cause the wilting of plants. These grubs may be removed by hand and destroyed. It is also useful to mix a little BHC and DDT (5%) dust with the soil. The aphid appears in winter and it can be controlled by spraying the plants with Malathion (10 cc in 10 liters of water).

Among the diseases, wilt and powdery mildew are important. The wilted plants should be uprooted and burnt as soon as they appear. Spraying with Karathane or Benlate (0.1%) is effective against powdery mildew.

ORCHIDS

Among the native flowers, orchids are most important and have been introduced into several countries from India. The Indian species of orchids have been used by orchid-breeders abroad for crossing with other species to produce some very attractive interspecific and intergeneric hybrids. The modern hybrids of orchid are used for cut-flowers which are the most expensive among all the flowers.

Several species of orchids are found growing wild in the forests of the Himalayas, particularly in the north-eastern region comprising Darjeeling, Sikkim, Bhutan, Nepal,

Shillong and the Assam hills, in the Western Ghats, Kodaikanal and some other areas. Orchids are also the natives of other parts of the world, e.g. Burma, Ceylon, Java, Singapore, Malaysia, Borneo, Thailand, Hawaii, New Guinea, South Africa, the USA, South America and Mexico.

Habit

In the tropics, most of the orchids are usually epiphytic, i.e. aerial plant growing on the branches of trees or on bushes, but not deriving nourishment from them. There are also terrestrial species which grow in the soil like other plants. A few species are also saprophytic and live on dead, decayed or dried animal or vegetable matter. They are, however, not cultivated, as it is difficult to provide them with the same conditions for their growth as are found in nature in which they thrive well.

Flower

The orchid flower is characterized by its three sepals, three petals and the column or gynostemium haiving the reproductive parts. Of the three petals, two are identical, whereas the third is highly modified and is the showiest part of the blossom, commonly known as the lip or labellum. The lip may be tubular or broadly expanded, with warts or protuberances and in a few cases it may be almost indistinguishable from the other petals. The flowers may be borne solitary, in sprays of inflorescences in clusters, spikes, racemes or panicles. The colours are vivid, solid, streaked, spotted or mottled.

Propagation

The orchids are propagated by dividing the clumps of psedo-bulbs or stems after flowering. They are also multiplied by dividing them into pieces, each having some roots, by cuttings (as in the case of Renanthera, Aerides and

Saccolabium), each cutting having some aerial roots. Orchids are also propagated from seeds (particularly the hybrids) in an artificial culture medium. However, the plants raised from seeds take at least three years to bloom. The modern technique of propagation of healthy plants is by meristem tissue-culture on artificial culture media. Nurseymen and orchid-breeders abroad use this technique for the rapid multiplication of orchids.

Cultivation

Orcjids thrive well only in places such as Assam, Darjeeling, Andhra Pradesh, Meghalaya, Tripura, Sikkim, Bhutan, Kodaikanal, Bangalore, Pune and Bombay. In the northern plains, most of the orchids, except a few do not grow well and usually die during summer. They can be successfully grown both indoors and outdoors. Generally, they thrive well in the fern-house or greenhouse or conservatory covered with creepers, and kept cool and moist by spraying water in the form of a mist on the plants during summer.

Of the numerous species of orchids, only a few can be successfully grown in the plains. The orchids that flower in the plains year after year are limited to the species Aerides multiflora (*A. affine*), Aerides odoratum, Cymbidium aloifolium, Phaius wallichii (terrestrial), Saccolabium guttatum, Dendrobium pierardii, Dendrobium moschatum, Pholidota imbricata and Vanda tessellata variety. unicolor (*V. roxburghii*) which are the natives of our country.

The epiphytic orchids are usually grown on hanging logs, or on small pieces of wood, with their roots covered with a thin layer of dried Sphagnum. Sometimes, they are also grown in small hanging earthen pots which are perforated on the sides for aeration and drainage. They are also grown in baskets. The potting-mixture contains five parts of small broken pieces of brick and one part of fern roots of bird's-nest or Polypodium.

A mixture containing equal parts of small broken pieces of husk, dried moss or peat, small dried pieces of bone and small pieces of charcoal can also be used. For terrestrial orchids, the potting-mixture contains equal parts of cowdung manure, leaf-mould, soil sand and crushed charcoal. The potting of plants should be done during November to February. They will bloom in March-April.

Diseases and Pests

Orchids are attacked by virus diseases. They are also attacked by insects such as thrips, red spiders and mealy bugs. The use of insecticides is effective in controlling these pests.

Species

There are hundreds of genera and thousands of species and varieties of orchids. A few very important species include Aerides odoratum, Calanthe vestita, Cattleya gigas, C. mendelii, Coelgyne cristata, Cymbidium giganteum, C. eburneum, Cypripedium insigne, Dendrobium densiflorum, D. draconis, D. aggregatum, D. fibriatum, D. nobile, Phaius wallachii (terrestrial), Phalaenopsis amabilis, Renanthera coccinea, Rhynchostylis retusa, Saccolabium giganteum, Paphiopedilum, and Vanda caerulea.

GLADIOLUS

With its majestic flower spikes with massive florets of brilliant colours, attractive shapes, varying sizes and excellent keeping quality, the gladiolus is ideal both for garden and for cut-flowers. It was brought into cultivation from its native habitat in South Africa, perhaps during the ancient Greek period. Its systematic improvement began only in the early years of the present century after the discovery of the Primulinus (*Gladiolus primulinus*), gladiolus growing wild near the Victoria Falls in South Africa.

The hybridization of Primulinus and a few other species, e.g. G. byzantinus, G. psittacinus, G. cardinalis, G. childsii, G. colvillii and G. gandavensis, resulted in the present-day gladiolus. It is undoubtedly the best bulbous flower in our country and ranks next only to tulips in Holland and other countries.

Uses

Gladiolus is excellent for growing in beds and pots and as herbaceous borders and for cut-flowers. As soon as the first floret has started opening and the others are showing colour, the flower spike is cut neatly with a sharp knife and is immediately placed in water. The other top florets will then open gradually and last for a long period.

Types and Varieties

The two most important types are the large-flowered varieties and the butterfly or the miniature gladioli. Both types are early, mid-season and late-flowering requiring 65-75 days to 100-120 days for flowering. The butterfly types have small spikes of various colours and in many cases with dark and attractively coloured throats. They are ideally suited for small gardens and flower arrangements. Besides, the primulinus and colvillei are two other types; the former have small florets borne on thinner spikes with a hooded top petal in each floret. Psittacinus hybrids are also popular in gardens.

The flowers of gladiolus may be large, medium or small, sometimes with petals ruffled, blotched or streaked. The colours range from white to near black, including pink, salmon, orange, red scarlet, maroon, yellow, greenish, purple, lilac, mauve, violet and several other shades. Some varieties are also dark or 'smoky'. There are also double-flowered varieties. Recently, scented gladiolus varieties have been evolved through the hybridization of Gladiolus with Acidenthera bicolor var. murielae. This intergenic hybrid is known as 'Gladenthera' and one such variety is 'Lucky Star'.

The important varieties are:'George Mazure', 'Fay', 'Gold Dust', 'Tunia's Yellow Triumph', 'Elizabeth the Queen', 'Jo Wagenear', 'Spic and Span', 'Spotlight', 'Dream Girl', 'Goeff Whiteman', 'Blaur Dominos', 'Polygoon', 'Banaras', 'Mary House', 'Snow Princess', 'Ratnas', 'Butterfly', 'Masholra Butterfly', 'Sylvia', 'Apple Blossom' and many others.

Cultivation

Gladiolus grows almost equally well in beds and pots. The beds should be dug out thoroughly and prepared finely. Before the final preparation, rotted cowdung manure, leaf-mould or compost about 5 to 6 kg and superphosphate 60 g per square metre may be added to the soil.

Gladiolus is generally planted from corms. It can also be grown from seeds, but the variety does not breed true in such cases. Propagation from seed is used for evolving new varieties through hybridization. The size of the spike and the flower depends upon the size of the corms planted. A medium-sized corm (10-12 cm), with a high crown is better than a larger and flatter corm. The corms are planted about 15-20 cm apart in rows spaced 30-45 cm apart. Sometimes, they may be planted in clumps of three or four, particularly for mass effect or as borders. The depth of planting corms is about 10-12 cm. Deeper planting being adopted in a lighter soil.

In the nortern plains, it is best planted from September to October and sometimes even earlier in August. Flowering takes place from December to April. In the hills, planting is done in March-April and it flowers from June to September. The time of flowering depends upon the variety and the time of planting. Successive planting of suitable varieties will ensure continued flowering over a longer period.

Manuring

Gladiolus grows best in a bed which has been manured previously for another crop. It does not require heavy manuring or excessive nitrogen as this practice tends to deteriorate the quality of the flower spikes and the keeping quality of corms. The plants respond better to organic manure than to artificial fertilizers. After 6 to 8 weeks of planting, the plants may be fed with liquid manure, about once a week. An application of light soot water can also be given once a month or once a fortnight. At the time of emergence of flower spikes, a dressing of about 2.5-cm-thick layer of leaf-mould is beneficial in producing better flowers with stronger spikes and better development of corms.

Watering

The plants may be watered weekly or fortnightly depending upon the weather. During winter frequent watering may not be needed.

Earthing and Staking

After about 6 to 8 weeks of planting, the plants may be earthed up. After the emergence of flower spikes, they may be staked, so that the spikes do not fall down when there is a strong wind. However, staking is not necessary when the plants are grown closely or in clumps, and it is not needed in the case of miniature, butterfly or primitive gladioli. Care should be taken not to injure the leaves at any stage as the injury is harmful to the plants.

Lifing and Storage of Corms

After flowering, when the leaves have turned yellow and have dried, the corms are lifted from the ground, preferably with a fork, and care is taken not to injure the corms while digging them out. After drying the corms thoroughly, they are put in paper bags, each with a few perforations for proper aeration or are wrapped up in sheets of newspapers, and stored in a cool and dry place or preferably in a cold storage room. The corms are best stored at 4.5o to 10oC and at 80-90 per cent relative humidity.

Diseases and Insect Pests

The commonest diseases are the Fusarium and Botrytis rots which cause the rotting of corms in the field and during storage. Botrytis rot occurs mainly in the hills. During storage, the corms are attacked by Aspergillus, Penicillium and other fungi, causing them to rot. In plants raised from disease-affected corms, the leaves turn prematurely yellow and flower spikes become stunted. It is not possible to control Fusarium and Botrytis rots easily with fungicides. However, to check the diseases, it is necessary not to grow gladiolus in the same beds every year, as the casual organisms are soil-borne.

The anti-biotic Aureofungin has been found to be effective in controlling the diseases of gladiolus. The corms should be planted after dipping them for one hour into 0-0.5 per cent Aureofungin solution. Two sprays of Aureofungin may be given to plants, one at the 6-leaf stage and the other 20-30 days before lifting the corms from the ground when the foliage is still green.

Thrips damage the leaves and petals by silvering them. They can be controlled by spraying the plants with Malathion (10 cc in 10 liters of water), once every fortnight or so. Caterpillars, which also injure the leaves and flower-buds may be effectively controlled by spraying the plants with DDT or BHC. It is useful to treat the corms with DDt or BHC (5%) dust and with Captan before storing them. The dusting of corms with naphthalene, about 30 g for every 100 corms and covering them with sheets of old newspapers or with gunny bags to retain the fumes may also be practiced before storing the corms.

BOUGAINVILLEA

The bougainvillea is a versatile ornamental plant for planting in both small and large gardens. It is highly prized for its beauty and utility. It is named after Louis Antoine de Bougainvillea, a French navigator. Being a native of tropical and subtropical South America, it is well adapted to our climatic conditions.

Species and Hybrids

The cultivars commonly grown in gardens belong to four botanical species and their interspecific hybrids, viz. Bougainvillea glabra, B. spectabilis, B. peruviana and B. buttiana. The species differ from one another in some characters such as growth, habit, leaf shape and size, colour of bracts and habit of flowering. Many cultivars have arisen as a result of natural crossing between two species, such as spectoglabra or glabra-peruviana hybrids, or as selections from seedling progenies. The variegated foliage types have evolved in nature as bud sports.

Varieties

There are hundreds of varieties of Bougainvillea in various attractive colours grown in gardens. The flowers of Bougainvillea are usually referred to the three coloured bracts, and real flowers are small and tubular, with a star-shaped apex that is often inconspicuous and attached to the centre of each bract. The bracts are white, light mauve, magenta, pink, terracotta, deep mauve, red, yellow, orange or lilac. The bracts may be small or large. Some of the outstanding varieties are 'Snow Queen' (white), 'Shubhra' (white), 'Sanderiana' (deep mauve), 'Trinidad' (pale mauve), 'Mrs. H.C. Buck (deep rose), 'Mary Palmer' (bicoloured deep rose and white flowers on the same plant), 'Louis Wathen' (orange), 'Enid Lancaster (yellow with shades of rose), 'Ladt Marry Baring' (Yellow), 'Mrs. Butt' (deep crimson), 'dr. R.R. Pal (brick red), 'Sonnet' (light rosy purple), 'Spring Festival' (medium magenta purple), 'Summer Time' (bright red), 'Partha' (young bracts orange changing to pink purple), 'Tomato Red' (terracotta), 'Sensation' (deep magenta), 'Blondie' (orange turning light rose), 'Isabel Greensmith' (coppery changing to rose), 'Bois due Rose' (biscuit colour), 'Begum Sikander' (white with rose edges) and 'Wajid Ali Shah' (rose). There are also multi-bracted cultivars like 'Mahara' (red), 'Roseville's Delight' (orange), 'Cherry Blossom' (white), 'Los Banos Beauty' (rose). A few important varieties having variegated leaves are 'Rao' (white and green bracts red), 'Thimma' (yellow and green, bracts bicoloured like 'Mary Palmer'), 'Louis Wathen' (variegated white and green, bracts orange) and 'Glabra' (variegated white magenta green). The variegated 'Mahara' - white and green and 'Archana' (variegated 'Roseville's Delight' yellow and green).

Uses

The bougainvillea can be used both as a shrub and as a climber. The shrub forms an attractive lawn specimen. It is also grown as a standard. A hedge of bougainvillea is quite common and colourful. It can also be trained on a tall tree, on the trunk of a dead tree or on a trellis, arch, pergola or screen. It is ideally suited for growing in large earthen pots, wooden tubs or cement pots.

Propagation

The bougainvillea is propagated from cutting, layering or budding. For budding, the commonest rootstock is the cultivar, 'Dr. R.R. Pal'. The best time for propagation from cutting and layering is during the rainy season (June-July). Seeds are used only for evolving new varieties. However, many varieties fail to set seeds but seed-setting is better in Bangalore, Mysore and Hyderabad than in Delhi and the Punjab.

Planting

The best time to plant bougainvillea is from July to September. The plant may be grown about 1.5-2.5 metres apart but in a hedge, a closer distance may be adopted. At the time of planting, about 8-10 kg of well-rotted cowdung manure may be added to the soil in each pit which is about 75 cm in diameter and 60 cm deep. The plants do not require much manuring. They should be pruned in June after they have finished their flowering to obtain better blooms in the next season, i.e. in winter (October-January) and in summer (March-June). After pruning, about 8-10 kg of cowdung manure may be applied to each plant, followed by copious watering.

The plant must receive full sunshine for good growth and abundance of flowering. The soil should be well-drained. During peak flowering, watering should be restricted; otherwise the flowers will fall off quickly under heavy watering. The plants usually need more frequent watering during summer than in winter. The plants are very hardy and, once established they grow successfully and flower profusely without much manuring or watering. The plants are almost free from disease and insect pests.

TUBEROSE

The tuberose is a native of Mexico. Its leaves are long, narrow, linear and grass-like. The flowering stalk emerges from the centre of the leaves. The flowers are tubular, single or double, white and highly fragrant. Usually, the single-flowered varieties are more fragrant than the double-flowered ones.

Tuberose can be successfully grown in pots and beds, on borders and as shrubbery. It is also used for cut-flowers which last long and are highly prized for their delightful fragrance. It is grown as a commercial crop for cut-flowers near Madurai in Tamil Nadu, in Bangalore (Karnataka) and also near Calcutta in West Bengal.

The bulbs are planted in September-October in the plains. The flowering takes place during summer and the rainy season (April to September) in the northern plains and in May-June in the hills. The plants thrive well in a well-draineds and sunny situation. They can also be grown in partial shade. After flowering, the flowering stalk should be cut down to encourage the production of more blooms in succession. The bulbs are left undisturbed in the ground and occasionally they are separated and replanted.

MARIGOLD

The African marigold is a native of Mexico and the French marigold is from Mexico and South America. Because of their being cultivated easily, wide adaptability to varying soil and climatic conditions, long duration of flowering and attractively coloured flowers of excellent keeping quality, the marigolds have become one of the most popular flowers in India. They are commonly grown in gardens, both in the urban and rural areas, and are cultivated commercially for use as cut-flowers, particularly for garlands. They can be successfully grown in pots too, and are used in mixed borders and beds. The dwarf varieties of French marigold are grown in window-boxes, hanging baskets, rockeries and as edgings. They are also ideal for growing in a newly planted shrubbery to provide colour and for a planting in blank spots in the garden. The dwarf French marigolds can also be grown effectively in drafts and along paths or driveways.

African Marigold

It is generally tall with large double globular flowers which are lemon, golden yellow, primrose, orange or bright yellow. There are also the nearest-to-white marigolds, though not pure white. There are also dwarf varieties having large double blooms. The varieties belong to two main types, namely the carnation-flowered with medium-sized carnation-like double flowers and the chrysanthemum-flowered, tall or dwarf, growing with fluffy and shaggy flower-heads like an incurved chrysanthemum. There are also the giant-flowered uniform and vigorous F1 hybrids, producing large globular fully double ruffled blooms in profusion and are ideally suited for cut flowers and garden display.

French Marigold

It is mostly dwarf, early-flowering and compact with dainty single or double blooms, borne freely and almost covering the entire plant. The flowers may be yellow, orange, golden yellow, primrose, mahogany, rusty red, tangerine or deep scarlet or of a combination of these colours. The flowers may be self-coloured, spotted, striped or botched.

Interspecific Hybrids

Interspecific hybrids between the diploid African marigold and the tetraploid French marigold have also been evolved. These interspecific hybrids are triploids, with intermediate characters, early flowering, medium tall, bushy with double flowers of delightful colour combinations of red and gold or pure yellow, orange, bright and rich mahogany scarlet.

The Single Signet

This is a dwarf type of marigold. It is bushy, with five lace-like foliage and covered with small single orange, yellow or lemon flowers. A dwarfer variety, 'Pumila', is very compact. The single signet marigold is ideal for edging and rock-gardens.

Cultivation

Marigolds can be successfully grown in different types of soil and climate. The French marigold grows best in a light soil, whereas the African marigold requires a rich, well-manured and moist soil. A well-drained soil and a sunny situation are essential for both types of marigolds. They can grow in almost all seasons, except in very cold winter, as they are susceptible to frost. The seeds are sown in May-June in nursery-beds or in shallow seed-pans or boxes. Seeds can also be sown in September-October and in February-March. In the hills, the sowing is done in March-April. After about a month of sowing, the seedlings are transplanted into beds or pots. The plants should be grown in a well-prepared soil which has received a liberal application of cowdung manure or farmyard manure.

The French marigolds should not be heavily manured; otherwise, they may produce excessive vegetative growth, resulting in poor flowering. As soon as the first flower-bud appears, the shoot is pinched to make the plants bushy and compact. In general, the marigolds are hardy and almost free from diseases and insect pests.

DEK

'Dek/'Dhrek' is an indigenous species commonly distributed throughout the country. In Panjab, the farmers grow this plant on field boundary or near tube-wells for shade or timber. It is fast growing, thin canopy and deciduous tree. It tolerates heavy lopping. The leaves are lopped for fodder. It is most suitable for growing on the boundary of the field. There are two types of Melia plants grown. The one with umbrella like spreading canopy is Melia compacta, 'Bakain' primarily grown for shade. The another type is avenue agroforestry species-Melia azedarach.

Climate and Soil Requirements

Melia grows well under dry conditions with rainfall of 60-100cm and temperature and 8 to 40°C. Young seedlings need protection from frost. It prefered well drained deep soils. It can grow on a variety of soils including saline and alkaline ones.

Silvicultural Practices

Nursery Raising

Propogated through seeds, cuttings and root suckers. Collect the ripe fruits in winter. These fruits are five seeded drupes and the seeds remain viable atleast for two years. Soak the seeds in water for 2-3 days before sowing to enhance the germination. The nursery is raised during February-March. Sow is soaked seeds in seed beds of one metre width and 4-5 metre length. Sow the seeds about 15-20cm apart. The seedlings start emerging after three weeks. Single fruit gives to 2-4 seedlings. Seedlings are spaced to 15 × 15cm by pricking out seedlings in nursery beds during July when these are 5-8cm in high. these are retained in the nursery beds for one year and are uprooted for planting out in the field.

Field Planting

Transplant are rooted dek seedlings during February or with earth ball during rainy season or in the beginning of spring where irrigation is available. Adopt spacing of 3 cm apart in boundary plantations. Irrigation and fertilization enhance the rate of growth.

Harvesting and Marketing

Harvest the trees in the beginning of winter. The wood is comparable to that of neem and is widely accepted in the market for furniture, doors and windows, packing cases etc. The leaves and fruits of Melia contain pest-repellant qualities.

KIKAR

'Kikar' or Babool' is a valuable source of fuel, timber, fodder and tinnin. The wood is used for making furniture, doors, windows, carts, trolleys, agricultural implements and other household articles. The pods and leaves contain digestible protein. The bark is a rich source of tannin. Kikar is grown in semi-arid and 'Kandi' regions.

Climate and Soil Requirements

It can be withstand extremes of temperature but it is frost tender when young. The tree can grow throughtout dry and hot regions of the country under average average annual rainfull of 2000 to 1200mm. It grows on a variety of soils ranging from sandy to loam. It can grow on soils upto pH 9.0.

Silvicultural Practices

Nursery raising :Collect seeds from superior, healthy and straight trees. Treat the seed in hot-water (80°C) for 10 minutes, followed by soaking in ordinary water for 24 hrs. Prepare raised beds of 1 × 5m size and sow at 0.5cm depth in lines which are 10-15cm apart. One square metre area needs 150-200g seed to produce about 1500-2000 seedlings. wooden boxes or earthen pots can also be used to raise the nursery on a limited scale. Sow the seed in February-March.

Prick out the seedlings when they are 5-10cm tall and transfer them in polythene bags of 22-10cm. Seed can be directly be sown in polythene bags.

Field Planting

Transplant the seedlings with earth balls during February to March where irrigation is a available, and during July-August in rainfed areas. July to August plantation has better survival. Dig pits of 50-50 cm well in advance. Mix 10-15g BHC 10% before filling each pits. fill the pits with 50 percent top soil and 50 percent Farm Yard Manure (FYM) and water them properly.

*Spacing:*For block plantation, plant at 3.0 × 3.0 m spacing. Thin out at the age of 5-6 years. For fuelwood plantation, 1.5 × 2.0m spacing is required with 5-6 years rotation. For boundaries and roadsides, two rows, 3m apart are planted.

*Tree-Crop Association:*To obtain better grain yield of wheat under kikar plantation (3 × 2m), PBW-299 (10.5 q/ acre) and PBW-175 (8.0 q/acre) varieties should be sown. Barley can also be grown under Kikar plantations.

Fertilization :Add 25 g third year of urea per tree during the second year and 50g during.

*Irrigation:*Irrigate at 10 days interval during the initial two years for better survival and growth.

Harvesting and Marketing

Nrmal rotation is about 20 years. With intensive cultivation, this can be advanced by 4-5 years. Average timber yield is 10-15 tonnes/year/ha. the trees are mostly sold standing to the local contractors. The price obtained depends on the age and size of the trees.

Seasoning

The greenwood contains about 70 percent moisture. Air drying to reduce the moisture to 12 percent level is achieved in 2-3 months for 25 mm thick planks. Solar kiln drying will reduce the time requirement to about 20 days.

Plant Protection Measures

The larvae bore into the stem and roots of the tree from mid-July to October. The growth of the attacked young tree is stopped and the tree ultimately dies. Avoid growing 'Ber', Casurarin a and Prosopis juliflora near Kikar plantations as they serve as alternate host of Babool borer.

NIM

'Nim'/'Neem' is one of the important indigenous tree as potential timber. Neem is also a source of fuelwood. The wood is relatively heavy with specific gravity varying from

0.56 to 0.85 (average 0.68). Timber is durable. Nim seed is a rich of source of non-edible oil (about 45%). During recent years, nim has caught the attention due to anti-insect, antifungal and antiviral activities of azadirachtin-an extract from neem seeds. Nim cake-blended urea adds to the efficiency of nitrogen use. The tree is widely distributed throughout sub-tropics and dry zones of western India and parts of Deccan. It grows well in Punjab. Climate and soil requirements It is widely distributed in dry and semi dry conditions and can tolerate long drought conditions. suitable rainfall is 450-1150mm, however, it can survive even below 130mm annual precipitation. Frost affect the growth of young neem trees. It can grow on variety of soils, including dry, stony, clay and shallow soils. It does not grow well on saline or water logged soils.

Silvicultural Practices

Nursery raising:Raise the nursery either in seed beds or in polybags. When raised in nursery beds, do prick out at 15 × 22 cm spacing or transfer in polybage filled with field soil and F. Y. M. (1:1). About one kg of seed is required to raise 100 plants. Fruits are collected in July-August when they are fully ripe. Extract the seeds with care and sow immediately.

Field planting:Transplant nim in the field at 3 × 3 metre spacing. Wider spacing (6m apart) is suggested where intercrops are to be raised. Transplant seedlings in August-September in pits (50 × 50 × 50 cm) filled with 50 percent top soil and 50 percent FYM. Plant the seedlings with earthen balls intact. Regular weeding and hoeing during the initial years is recommended. It is nor readily browsed by cattle and goats. Neverthless, protection is needed from damage by animals.

Fertilization and irrigation:Fertilization during the initial 2-3 years helps better establishment and growth. Irrigation is applied 10 days interval during dry spell in summer months. Whereas during winter it may be at fortnight interval.

Harvesting and marketing:Harvest the trees in late winter. The tree on an average puts girth increment of 2.3-3.0 cm in unirrigated and 8 cm per year in irrigated is plantations. The total volume including fuelwood obtained to the extent of 120-170m*/ha in 8-10 years. The wood is used as timber and fuel. Timber is used in housebuilding, ploughs and agricultural implements, making boards, panels, and toys. When large plantations are raised, collect the seeds every year and supply to the concerned orgations for oil extraction.

PAHARI KIKAR

Pahari Kikar (Mesquite) is native the Central America and the northern south America. It is a thorny deciduous, larged crowned and deep rooted bush or medium sized tree. The tree has been planted in many arid zones of the world. The species is widely propogated, mainly for stabilizing dunes. It is excellent for firewood and makes superior charcoal.

Climate and Soil Requirement

Mesquite grows in very warm and dry climate from sea level to 1500m altitude. It is least affected by drought. It prefers sandy soils. It grows well on rocky terrain provided that root growth is not impeded. Can tolerate temporary water logged conditions and grows well in salt affected 'Kallar soils'.

Silvicultural Practices Nursery Raising

Collect the seeds in May-June and scarify mechanically or treat with 20% sulphuric acid for one hour. Wash thoroughly and keep in water for 24 hours. Sow the seeds in polythene bags. Germination of treated seeds is 80-90 percent.

Field Planting

Transplant the seedlings in august-September or in February-March depending upon the availability of water. dig the pits of 30 × 30 × 30 cm size at distance of 1.5 × 1.5 m. Add 5 kg of well rotten FYM and 10-15g BHC 10% with soil and fill the pits. Irrigate the pits. Prefer planting of this tree on the waste lands specifically on salt affected soils.

Harvesting and Marketing

Early harvesting is generally done to meet local fuel requirement. The plants, however, when grown for small timber are retained for 10-15 years. Wood makes good fuel, charcoal, fence post and light timber.

POPLAR

Poplar is among the first growing industrial soft woods which can be grown well as pure as well in association with agricultural crops. The wood being light, homogeneous, without any taste or flavour is in much demand for paper and pulp, plywood, match sticks, packing cases, sports goods, light constructional timber and furniture.

Climate and Soil Requirements

Poplar grows well in the subtropical climate where the temperature extremes are not too severe. During summer period adequate irrigation facilities are required. The spring must be free of late frost. The subtable soils with texture ranging from sandy loam to loam and pH between 6.5-8.5.

Silvicultural Practices

Nursery Raising

Propogate poplar through stem cuttings of 20-25 cm length and 2-3 cm diameter prepared from one year old plants.

Soak the cuttings in fresh water for 24 hours before planting. Plant the cutting in the beds (5 × 15 m) prepared after 8-12 tons FYM and 40 kg BHC 10% per acre. Apply P_2O_5 form 40-80 kg and K_2O from 20-40 kg/acre depending upon the soil type. And 100 kg urea/acre in two split doses during rainy season. Dip the cuttings for 24 hours in fresh water. Treat the cuttings before planting with 0.5 percent emulsion of durmet and 0.5 percent solution of Emissan-6 wetable powder in water. Dip the cuttings for 10 minutes in each solution. Plant the cuttings at 50 × 50 cm apart during the first fortnight of February. Apply light irrigation at 7-10 days interval. Hoeing, weeding and debugging are required in nursery. Plants are ready for transplanting after one year.

Field Planting

Transplant bare-rooted plants in mid January to mid-Feburary in channels. Dig the pits of 15-20 cm diameter with the help of auger of "Boky". The depth of pit should be 75 cm in clayey soil and 100 cm in sandy loam soil. Soak the plants for about 48 hours in running fresh water before planting. Treat the lower 1 cm of the plants with 0.25 percent aldrix solution for about 10 minutes and then with 0.15 percent solutions of Emissan-6 for about 20 minutes. After planting, the pits should be filed with top soil and FYM (1:1) mixed with 10-20 g BHC 10% and 50 g P_2O_5.

Spacing

In block plantations, plant the poplars at 5 × 4 m apart and in single line on field boundary at 3 m apart in North-south (N-S) direction.

Tree-Crop Association

It is desirable to grow field crops as intercrops in block plantation of poplar trees. Intercultivation of crops in polar plantation is a bonus in addition to better performance of the tree growth. All the rabi and Kharif crops can be grown during the first three years except paddy. Intercultivation of sugarcane for the first two years is more profitable.However, poplar being a deciduous tree, rabi crops like wheat, oats, berseem, sarson, vegetable etc. can be raised throughout the tree rotation. For obtaining better grain yield of wheat about 12 q/acre through intercultivation in poplar plantations more than three years old, sow wheat varieties namely PBW-299, PBW-175, PBW-154, WH-542, and HD-2329 in the first week of November. Delayed sowing of these varieties causes decline in the yield. Under late sown conditions for securing best grain yield prefer PBW-226 and HD-2285.

Irrigation

Irrigate immediately after planting and keep the poplar channels moist for first two years. Subsequently, irrigate fortnightly during October-February and at 7-10 days interval during March to June.

Plant Protection Measures

Protecting from pests: The insecticides recommended for the control of various should be various insects should be sprayed using 200-250 litres of spray material per acre with manually operated rocking sprayer (Gatore) or with a long arm tractor sprayer. The quantity of spray material to be used will depend upon size of trees. However, actual amount of insecticide recommended should not be reduced.

*Poplar leaf defoliator:*The caterpillars of these moths feed on leaves. The onfestation starts in March-April with peak period of activity from July to September. control the insect by collecting and destroying infested leaves. Spray monocrotophos 36 SL @ 600 ml (Monocil/Monolik/Nuvacron/corophos/Luphos) per acre.

*Leaf Webber:*The young larvae scrap the leaf surface along the veins and feed on epidermis of leaves by webbing 2 and 3 leaves with silken threads. The pest is active from April to November with peak period from July to October. Adopt the same control measures as given under leaf defoliators.

*Bark eating caterpillar :*Noctumal feeding larvae make L shaped holes and wet silken threads entangled with faecal pallets. Pest is active throughout the year. Prunesecerely infested branches and spray suspension of 100 g Carbaryl 50 WP (Sevin/Hexavin) in 10 litres of water during September to October at feeding sites.

*Case Worm :*The pest is active throughout the year. The caterpillars feed on bark from December to March, on leaf buds during March to April and on leaves from April to November. Spray carbaryl 50 WP (Sevin/Hexavin) @ 1 kg/acre.

*Leaf hopper :*The leaf hoppers are active from April to November with peak period of their activity from July to October. Spray oxydemeton-methyl 25 EC (Metasystox) @ 300 ml or dimethoate 30 EC @ 250 ml per acre.

Diseases

*Rot of cuttings:*Black dots appears on cutting at ground level and decay of bark takes place. dip the cuttings for 15 minutes in 0.5 per cent solution of Emisan-6 before planting.

*Leaf Spots:*Brown to dark brown leaf spots af variable sizes appear on leaves. Severe infections lead to premature defoliation. Spray the crop with 0.25 percent copper chloride (Blitox 50) or Indofil M-45 at 15-20 days interval starting with the first rain. Give 2-3 sprays.

*Pink disease :*Gridling of branches in young plants leads to death of parts. The height of tree is stopped due to repeated death of the leaders. Pink to salmon coloured mycelial growth appears on branches. Use resistant varieties or 2-3 prophylactic spray of Bordeaux mixture during 2-4 years of age at the beginning or summer monsoon.

*Sunsclad canker:*Bark is killed due to insolation by hear and canker develops on the southern side of the stem. Protect from insolation and other injuries by white eashing the main stem upto 2 metres from ground level.

*Bark Bursts and Canker:*Water oozes out through the wounds resulting in cankers. Avoid injury and high water table sites. Clean the wounds and apply Bordeaux paste and Emisan-6.

Harvesting and Marketing

The plants are ready for harvest when they attain the grith of about 100 cm. Marketing is done as standing trees or after felling and cutting into logs. The wood is utilized by Western India Match Company (WIMCO) and plywood, sports goods, and packing cases industries.

SAFEDA

'Safeda' is the most widely exploited exotic genus in the world. With over 630 species, originated primarily in Australia, the genus offers a very wide range of plant types suitable for different soils and climatic zones. Fast growth, straightness, self pruning and wide utility of wood are some of the main features of 'Safeda'. Presently, the wood is being used for furniture, doors, windows, crates, packing cases, fuel, pulpwood and even as a source of rayon. The tree is used as wind break especially for the protection of fruit crops. there is potential scope for 'Safeda as a timber tree' to attain profitable status in the years to follow.

Climate and Soil Requirements

The tree withstands high range of temperature and can

grow wide range of annual rainfall (20-125 cm). It grows well on deep soil with substantial sub siol moisture having pH between 6.5 to 9.5.

Propagation

Seedlings can be procured from the State Forest Department. Seeds can also be collected from healthy and biggest trees. sow the seed during February-March or September-October on raised beds either by broadcasting or in lines 10 cm in apart at a rate of 20 g/m2. cover them lightly with soil. covers the beds with a thatch and sprinkler water frequently to keep the upper soil layer moist. Transfer the seedlings, when these are 8-10 cm in height, in perforated polythene bags (22 × 15 cm) filled with a mixture of soil, sand and FYM (1:1:1). Keep the bags under partial shade in sunken beds flooded with water. Seedling become ready for field planting in about 4 months.

Transplanting

Transplant the seedlings during August-September or February-March in pits (30 × 30 × 30cm). The pits should be filled with upper soil and FYM (1:1) mixed with 10-15g BHC 10%. Plant the seedlings in the centre of the pit after gently removing the polythene bags. Care should be taken that the earth ball and roots may not get damaged. Water the plants immediately. Raise the boundary plantations at 2-3 m. apart and block plantation is done at 3.0 × 1.5m.

Tree-crop Association

'Safeda' is commonly planted on the field boundaries. It may have an adverse effect on the adjoining field crops if not properly managed. following measures help the proper management.

(i) Keep the boundary plantation in North-South row direction to avoid continuous shading on the northern aspect in winter (rabi season).

(ii) The distance between trees should be 2-3 meters. Avoid double row plantation.

(iii) Seperate agronomic management, particularly sowing time and irrigation schedule, is required for the crops grown in a 10-15 metre wide strip running along with the boundary plantation. Avoid additional application of fertilizer crops

(iv) Prefer folder (guinea grass, napier bajra, berseem, oats, bajra, jowar, cowpea, etc.) rather than grain crops along with the tree line.

(v) Keep water channel along with the tree line.

To obtain better fodder yields prefer intercultivation of Guinea grass, Cowpea and Oats in eucalyptus plantation. In addition to the Bajra and Sorghum can also grown successfully. During the initial one to two years sow smaller crops like pulses, berseem etc. to avoid suppressing effect of the taller crops.

Apply at each plant, 50g and 100g of urea in two split doses during second and third year, respectively. After third year no fertilizer is required.

Irrigation

Irrigate twice a month in summer and once a month in winter upto the age of 3 years.

Traning and pruning

'Safeda' is a self pruner leader. In case there are two or three leader shoots, retain only one healthy and straight one. Cut heavy branches, if any.

Thinning

Remove weak, diseased, suppressed and dead plants after 3-4 years and maintain 700 trees/acre for pulp and pole. If the plantation is maintained for timber then again remove, damaged and weak trees at the age of 7-8 years and keep 400 healthy trees/acre for a rotation age of 12-15 years.

Harvesting and Marketing

For good marketing harvest the trees for timber after 12-15 years. for paper pulp, fuelwood and props fell the trees at 7-8 years age. Fell the trees from November to February or during rainy season. Dry the logs in shade.

'Safeda' is sold either by weight or by volume. Price varies with the thickness of trees. Contact the nearest Forest Office or Punjab State Forest Development Corporation or the Department of Forestry Natural Resources, Punjab Agricultural University for estimating the value, and disposal of the trees. Paper mills, plywood and packing case industries are the main wood consumers.

Diseases

Cylindroclaldium seedling blight: The disease occurs in nursery and out plantings. Infection on leaves results in defoliation and wilting of branches. Initial symptoms appear as dot-like spots which later become greyish black. Incidence of disease declines in dry weather. Spray of indofil M-45 @ 0.5 per cent or Bavistin @ 0.1 per cent is done on the leaves.

Dumping off of seedings: Rotting of seedling at soil level. Avoid high density excessive irrigation and undercomposed FYM. Sterilization of seed beds with Formalin (250 ml in 4 litres of water per sq. metre) should be done. Three soil renches of Bavistin (0.1%) or Indofil M-45 (0.2%) or Emisan-6 (0.5%) at intervals of 4-5 days are also effective.

Gummosis (Physiological disease): Swelling and splitting of bark and exudation of shining golden brown viscous liquid takes place which hardens as brown or reddish lumps on tree bark. Avoid injury to the tree.

SAGWAN

Climate and Soil Requirements

Teak thrives best in a fairly moist, warm tropical climate with annual rainfall 125-400 cm, and temperature from 10-40°C. Grown on a variety of soils, including laterite, gravelly, acidic and alkaline soils. Well drained sandy loam soil is best. It requires good drainage and cannot tolerate water stagnation.

Propagation

Collect the seeds in February and March. Use large and medium size seeds for nursey development. Give nay one of the following pretreatments to the seeds to soften the hard coat.

(a) Soak in water for 24 hours and dry in hor sun for 24 hours alternately for 2-3 weeks.

(b) Acid treatment with concentrated sulphuric acid for 48 hours followed by thorough rinsing in water.

(c) Dumping in moist FYM for 25-30 days.

(d) Scorching in light fire of leaves and grass.

Nursery Raising

Sow the seeds on standard raised beds, 12m × 1.5m × 20m in March-April. Covers the beds with thatch and water daily till the germination completes within 20-30 days. the sedling attain stumpable size in one year.

Field Planting

One year okd seedlings/stumps are planted in July-august into 50 × 50 × 50cm pits at the distance of 2 × 2m.

Tending

Train the plants the boundary to have clear and straight

bole. In block plantation, normal weeding and irrigation required for the first 2-3 years. First thinning is done when the crop is 8-9m high. when the height is 10-15m, remove alternate plants. Subsequently thinning of diseased/ deformed trees should be done. Pruning before the first thinning is essential.

Harvesting and Marketing

It attians the exploitable diameter in 60-70 years. In good and productive site, it is ready in 40-50 years. It is sold as standing tree or as logs converted into variable sizes. Primary use is timber. Used in construction, furniture making, general carpentry and as a raw material for the timber industry.

Pests and Diseases

Teak is subjected to various insect attacks. Defoliators and stem borers are common in pure crops. However, fungal damage is very little, the only exception is uncinula tectonae on leaves.

SUBABUL

Subabul is a miracle tree as it provides fodder, fuel, pulpwood and timber. The forage is highly palatable, digestible and protein rich. Subabul helps to enrich soil and aid neighbouring plants. Under good soil and moisture conditions, one hectare os subabul bushes pruned and mowed in soil can add upto 500 kg of nitrogen per year and thus is a good source of biofertilizer. the wood is good for fuel, charcoal, pole, and pulp.

Climate and Soil Requirements

It grows in best areas having 500-3000mm annual rainfall. The plants can survive in dry seasons lasting for 8 months. It grows very well on neutral of alkaline soils. It can thrive on soils with texture varying from rocky to heavy clay. It is also well adapted to the poor soils and dryland conditions of 'Kandi' area.

Varieties

K-8:Vigorous in growth reaching height of 20m. The angle of branches is narrow. thrives well under rainfed conditions. Biomass production is 30-40 tonnes/ha.

Cunninigham:Spreading branches having having more leafy fraction. The mimosine content is comparatively low. Biomass production is 30-35 tonnes/ha.

Silvicultural Practices

Nursery Raising

Nursery is grown on raised beds or in polythene bags during the beginning of rainy season. Seed can be sown directly in the field. for direct sowing, 5 kg of seed/ha is used. For raised beds, 25g of seed is required to raise nursery in 1 sq. metre area. Seeds should not be sown deeper than 1-2 cm. Germination occurs within 6-8 days. Germination can be hastened by hot water (80°C) treatment and soaking the seed in water for 24 hours before sowing.

Field Planting

Direct seed sowing is done in rows 75 cm apart for forage production accomodating about 1,00,000 to 1,20,000 plants per hectare. for small diameter wood, plant 10,000 or more trees per hectare. Do thinning so that the remaining trees can grow trunks large enough for fence posts, pulpwood and for lumber.

Tree-Crop Association

Lopping/pruning of Subabul is essential factor to grow agricultural crops as intercrop. Otherwise it reduces crop yields considerably by casting shade effect due to the fast growth. The lopped subabul hedges provides lush green nutririous fodder throughtout the year. In addition to soil conservation of benefits in subabul alleys (1.5 m stem height) planted at 5 × 2 m distance, sorghum, cowpea, bajra and oats can be grown for fodder. In addition to fodder yield, subabul alleys produce about 31.4 and 17.7 q/acre/annum fodder and dry fuel, respectively. In between subabul alleys (rows). wheat, barely ans sarson crops be grown successfully. These crops can grow better in alleys (rows) than sole crop. In 'Kandi' region, bajra-oats rotation in between subabul line (3-4m × 1-1.5m apart) is beneficial.

Fertilization

Apply basal dose of nitrogen (30kg/ha), and phosphorous (60kg/ha) at the time of planting.

Irrigation

Irrigate frequently when grown exclusively for fodder. For pole/timber plantation, irrigation is required for the first three years only.

Hoeing and weed control

Weeds are a major cause of failure or slow establishment of subabul plants. Regular weeding (10 to 15 days interval) until plants are 1-2 metre tall gives best results.

Harvesting and Marketing

The crop harvested after every 50 days gives forage yield. for timber, it is cut at 12-15 year rotation.

TAHLI

Tahli/Shisham', a native of Indo-gangatic plains, is a multipurpose and moderately fast growing tree. Every part of this tree is used for economic purposes. Heartwood provides excellent timber and leaves provide a nutrious fodder during lean period. It is winter deciduous, and fixes atmospheric nitrogen to enrich the soil. It is our state tree and is thus found scattered along roads, canals, railway lines, and on agricultural farms.

Climate and soil requirement

Grows well below 1000m elevation. 'Tahli' prefers temperature from 6-40°C and rainfall of 20-125 cm/annum. It grows well on alluvial soils and prefers sandy loam soil where deep penetration of roots is possible. It can also grow in slightly saline soils.

Silvicultural Practices

Nursery Raising

'Tahli' is propogated from seeds and suckers. collect the ripe pods during december-January from healthy and straight trees. Sow the seeds or cut pod pieces containing one seed each on raised beds or in containers or directly in the field. Prepare raised beds from well worked soil mixed with 10-16 quintal FYM and 40-60 kg urea per acre. Loam soil is most suitable for nursery raising. Soak the pods in cold water for 48 hours. Sow the seed 1.0-1.5 cm deep either in the second half of February or in July-August. For one acre nursery, 2.0-3.5 kg of pods are sufficient which produce about 60,000 seedlings. Irrigate the beds regularly. Avoid water-logging.

Germination starts after 10-15 days and is completed in about 3 weeks. Thin the seedlings to give space of 15 × 10 cm when they are 5-10 cm in height. Regular irrigation and hand weeding are necessary to raise healthy seedlings. The seedlings are intolerant shade. Nine months old seedlings become ready for preparing root-shoot cuttings. Chop off to about 7.5 cm shoot and 22.5 cm main root. The entire plants can also be planted.

Field planting

Planting is done in pits, which are prepared in advance at 2 × 2 m in the field. remove every alternative plant during 6th year. Remove the dead, dying, misshaped, diseased trees and maintain best 100 trees per acre. These tree should be evenly distributed over the area.

Irrigation

Irrigate at the time of planting and subsequently for another 2-3 years. Once established, will not require much irrigation.

Seasoning

Air seasoning of logs for atleast six months is necessary to overcome the wraping and twisting. The timber dries readily, if stacked carefully. Stacking should be under shade on levelled ground and every planke should be well balanced. Stack the planks transversely so that sufficient air passes through them. Kiln seasoning takes less time. (10-15 days) and is now available in many cities like Phillaur, Kartarpur, Sirhind, Hoshiarpur, etc.

Diseases

Root rot:The diseases prevalent in both heavy and light textured soils, but the spread is quicker in light soils and trees are killed in a short period. Affected trees show die back symptoms, leaves become pale and fall off. At later stage, reddish-brown coloured fruit bodies of the pathogen appear at the base of tree trunk during rains. roots get decayed and become spongy white and very fragile. To control, raise plantations through seedling. Injuries to the roots and basal stem should be avoided. Infected trees should be isolated from healthy trees by digging 5' × 2' × 1' trenches around them. Drench the roots with Bavistin or Vitavix @ 0.1 percent (4.0g in 40 litres of water per tree).

Harvesting and Marketing

Fell the trees at the age about 40-50 years age for timber production. Its heartwood is brown in colour, very hard, strong, durable and possesses beautigul natural grains. It is used for making cabinets, furniture, decorative ply, and the constructional timber. Being an excellent timber there is no marketing problem for Tahli logs. The price in the market varies with length of the log and quantity of heartwood present. The mature trees are usually sold locally to contractors and saw mills.

TOOT

'Toot'/'Mulberry' is a common tree found all over India. Its native home is Japan from where it has spread to other countries. It is a medium sized tree with deciduous habit.

Varieties

M. alba, M. nigra and M. rubra are the species which are cultivated. M. alba is mainly cultivated for the silk-worms whereas M. nigra is cultivated for fruit and M. rubra is cultivated for timber as well as for fruits. The wood is yellowish-brown, of good quality and is used for furniture, turney, hockey sticks and boats. Twigs are used for baskets and as fuelwood.

Climate and Soil Requirement

Toot can grow in temperature, sub-tropical and tropical regions. It thrives well both under temperature as well as sub-tropical conditions, and is not affected by drought or freezing. It can be grown in any type of soil and performs better in clay-loam.

Silvicultural Practices

Nursery Raising

The common method of propagation is by cuttings. Prepare

stem cuttings about 23 to 30 cm long having 3 to 5 buds from the mature shoots in December or January. Plant the cutting 10 to 15 cm apart in the flat nursery beds in rows that are 30cm apart. Water these cuttings lightly. Usually they start sprouting by March and are ready for transplanting during the rainy season.

Field Planting

For block plantations, dig the pits of 50 × 50 × 50 cm size at a distance of 6 × 4 metres during november-December and keep open for about 20 days. Mix 2 to 3 baskets of well rotten FYM and 10-15 gm of B.H.C 10% with the soil and fill pits back. Water the pits so that the soil may settle well. Transplant the seedlings later when the soil is in "Wattar" condition. For sericulture, accomodate 1600-2000 trees per acre. For boundary plantation or as wind break, tree to tree distance is 2-3m.

Pruning

Prune all shoots upto height of 2-3m from the ground level in December-January. It helps to form a clear bole of the tree. Prune the plants raised for sericulture heavily even upto ground surface. It helps to produce big sized leaves.

Fertilization

Manure the trees which are grown for sericulture with pond mud at the rate of 8 to 12 metric tonnes per acre before the start of the leaf plucking period. Mix the manure with soil by ploughing. Do this type of manuring once in a 3 to 4 years. The plants are headed back to ground level after 10 to 12 years. Allow the new shoots which arise from the stumps to grow for 4 to 5 years.

Irrigation

Irrigate for better establishemnt. Give one or two irrigations during spring for better sprouting of the plant. Irrigate during summer after 15 days. No irrigation is needed after October.

Harvesting and Utilization

Twigs are used to make baskets and its wood in sports goods. the most important use of mulbery lies in rearing of silk worms.

REFERENCES

Dole, J.M. and H.F. Wilkins, *Floriculture principles and species*, 2 nd Edition. Pearson Education, Inc., Upper Saddle River, N.J. Available in Branson Library, 2005.

Larson, R.A., *Introduction to floriculture*, Academic Press, N.Y. [Branson Library], 1992.

Seeley, J.G., *Advancements in Commercial Floriculture*, Southern Florist and Nurseryman, 1982.

Prasad S. and KumarU., *Commercial Floriculture*, Reprint. Jodhpur, Agrobios, 2003.

Shanmugam, V., *Commercial Flower Cultivation*, New Delhi, Mohit Publications, 2007.

Singh, A.K., *Flower Crops : Cultivation and Management*, New Delhi, New India, 2006.

7

Trends in Commercial Floriculture

Farming flowers and foliage employs special aspects of floriculture, such as spacing, training and pruning plants for optimal flower harvest; and post-harvest treatment such as chemical treatments, storage, preservation and packaging. The production and marketing of flowers can be a profitable undertaking for small-scale growers. Flowers can be grown by themselves, or can be rotated with vegetables.

With our changing life styles and increased urban affluence, floriculture has assumed a definite commercial status in recent times and during the past three decades particularly. Appreciation of the potential of commercial floriculture has resulted in the blossoming of this field into a viable agri-business option. Availability of natural resources like diverse agro-climatic conditions permit production of a wide range of temperate and tropical flowers, almost all through the year in some part of the country or other. Improved communication facilities have increased their availability in every part of the country.

In India, as commercial floriculture is an activity which has assumed importance only in recent times, there are not many large farms engaged in organised floriculture. In most part of the country flower growing is carried out on small holdings, mainly as a part of the regular agriculture systems. The estimated area under flower growing in the country is about 65,000 hectares. The major flower growing states are Karnataka, Tamil Nadu and Andhra Pradesh in the South, West Bengal in the East, Maharashtra in the West and Rajasthan, Delhi and Haryana in the North. It must, however, be mentioned that it is extremely difficult to compute the statistics of area in view of the very small sizes of holdings, which very often go unreported. This perhaps would be the reason for unrealistically small areas reported for floriculturally active states like Maharashtra, Uttar Pradesh and Madhya Pradesh.

More than two thirds of this large area is devoted for production of traditional flowers, which are marketed loose e.g. marigold, jasmine, chrysanthemum, aster, crossandra, tuberose etc. The area under cut flower crops used for bouquets, arrangements etc. has grown in recent years, with growing affluence and people's interest in using flowers as gifts. The major flowers in this category are rose, gladiolus, tuberose, carnation, orchids and more recently liliums, gerbera, chrysanthemum, gypsophila etc.

The production of flowers is estimated to be nearly 300,000 metric tonnes of loose flowers and over 500 million cut flowers with stem. In the case of production also, the estimates could be at variance from the actual figures as some of the flowers like rose, chrysanthemum, and tuberose are used both as loose flowers and with stem. Almost all of the area reported here is under open field cultivation of flowers. Protected cultivation of flowers has been taken up only in recent years for production of cut flowers for exports. The estimated area in production is about 200 hectares, which is likely to increase to over 500 hectares by the year 2000. Recognising the potential for low cost production for export, in view of cheap land, labour and other resources, several export oriented units are being set up in the country. These projects, located in clusters around Pune in the West, Bangalore and Hyderabad in the South, and Delhi in the North, are coming up in technical collaboration with expertise mainly from Holland and Israel. More than 90 percent of these units are for rose production, on an average size of 3-hectare farm, while some projects for orchid, anthurium, gladiolus and carnation are also being set up. Nearly one third of over 200 proposed projects, have already commenced production and export.

SELECTION OF PLANTS

There are annuals, perennials, grasses, woody shrubs, trees and vines all suitable for use as cut flowers or other cut plant material. Flowers can be grown specifically for fresh use, dried use, for parts other than flowers such as seed pods, or any combination of the preceding. Start with species or varieties proven to grow in your area - Virginia's climates are as varied as its geography. Introduce other species on a trial basis before committing to large-scale plantings. An interesting industry trend is toward smaller acreage for cut flower production: 0.5 to 2 acres. Overproduction can be a problem. Most local markets can't accommodate the volume produced on 10 acres of cut flowers, no matter how beautiful they are!

Annuals will produce in the year planted and most are easily started from seed. Annual seed is readily available and lower in cost compared to perennials and bulbs. Tender or half-hardy perennials can also be grown as annuals. The choices for perennials are endless. Some traditional perennial crops that bring a good price include peonies (both fresh and dried), lily of the valley, and calla lilies.

Garden-type roses with fragrance are a possibility for the specialty florist market-also the heirloom roses such as cabbage and bourbons could fetch top dollar during the wedding season. Woody species can be used to extend production time to include early spring flowers (either naturally or on forced branches such as pussy willow or flowering apricot), summer flowers, late season berries, or ornamental stems such as the corkscrew willow harvested in the fall and winter. Relatively low maintenance and long-term production are benefits of including "woodies" in your field-grown cuts programme. Do keep in mind that it may take a few years for some species to reach marketable harvest size. Both wholesalers and florists want to be the first with unusual or improved varieties. If you have the space, consider a larger product mix. A wide variety of cuts allows you to service a greater percentage of the needs of your customers. Consistently test new crops. The ability to provide the floral wholesaler or retailer with unusual material will make you invaluable. Most wholesalers or retailers will pay for samples of a new variety. In the same vein, be aware of changes in consumer style and colour preferences.

METHODS OF FLOWER PRODUCTION

The ideal site for cut flower production is in full sun with wind protection, an irrigation source, and easy access. Raised beds are virtually required to ensure proper drainage (unless the soil is very sandy). Beds can be any length. Keep in mind that beds should be narrow enough for you, your employees, and/or customers to reach comfortably to the middle of the bed. A bed height of 4 to 8 inches should suffice, and drainage can be enhanced by burying drain lines 12 inches below the beds. Break up the hardpan or clay layer at least 1.5 to 2 feet deep with a subsoiler.

Proper preparation of the beds is essential to optimum growth and yield. Amend beds with plenty of organic material. Well-composted horse or cow manure is excellent. Be sure it has aged to the point where weed seeds are no longer viable. Incorporate lime, superphosphate, and other nutrients as recommended by the results of your soil test. If any of the beds are to lay fallow for a length of time, such as over winter, cover crops are excellent for increasing the organic matter content of the soil and reducing erosion. Recommended fall and winter cover crops for Virginia include hairy vetch, rye, barley, and crimson clover; or better yet, a combination of two or more of these crops. Summer cover crops can include soybeans, cowpeas, buckwheat, and sesame. Mulch has its proponents and detractors. Some growers insist on it for weed suppression and moisture retention. Others prefer to till between rows to control weeds. Because of the large variety of species grown as cut flowers, no single herbicide can be recommended.

Planting Material

Plant material can be acquired from a number of sources. Annuals can be direct sown into the beds or transplanted as plugs. Some annual species should be seeded in succession to continually produce as long as the growing season allows. Perennials are best started from plugs, whether seed or vegetatively propagated. There are many commercial sources for large plugs and starter plant liners. If perennial species are planted out in the spring, be sure they have been vernalised if that particular species requires it in order to flower that season. The requirement of planting material to cater to the large area under flower crops, is largely met from domestic production. Since efforts to set up large commercial farms generally suffered due to lack of quality planting material in sufficient quantities, this aspect has received greater attention in recent years in the breeding centres, which are producing sufficient quantity of planting material. Most of the nurseries propagating planting material are in the private sector. In the absence of any mechanism to register nurseries, it is very difficult to ascertain their exact number, but at a very conservative estimate there are more than 100,000 nurseries, spread out all over the country, producing seeds and other planting materials for flower growers.

The states with larger numbers of nurseries include Maharashtra, West Bengal, Karnataka and Tamil Nadu. Most of the nurseries are small, with little or no improved facilities like mist propagation unit, green houses/net houses etc. For meeting the demand of flower seeds, several large seed companies have production units in Punjab, Himachal Pradesh and Jammu & Kashmir in the North, Karnataka in the South and West Bengal in the East. A few of the leading multinational seed companies have tied up with local seed companies or producers for custom production of seeds of their varieties. In the case of bulbous plants, most of the planting material is produced in the north eastern hilly regions of West Bengal and Sikkim, though for some crops, it is also produced in hilly regions of northern India. The introduction of a revised seed policy by the government of India in 1989 has enabled unrestricted introduction of many new and superior varieties into the country, increasing the variety in the floral basket.

Tissue culture has, in recent years, been recognised as an important tool in agriculture development. With its diverse climatic zones and qualified manpower, India is well placed to exploit the benefit of tissue culture based applications to floriculture crops. Most popular application of tissue culture has been micropropagation using *in vitro* technique for mass multiplication of planting material. Tissue culture plants of ornamentals have found ready acceptance by the commercial growers and their production increased significantly from 130 million plants in 1985-86 to 680 million in 1994-95. At present 30 commercial tissue culture units with annual capacities of 0.5 to 15 million plants each are in operation, resulting in total capacity of about 110 million plants.

Spacing

Optimal spacing varies between species. What seems adequate spacing for a row of perennials the first year may result in overcrowding the following year. Dense spacing can lead to higher incidence of disease as air circulation is limited. Conversely, too much space is an invitation to weeds and reduces yield per square foot. Note that dense planting often encourages longer stems. The spacing of annuals varies by species, ranging from 4 to 6 inch centers to 1 foot x 1 foot. Be sure to thin rows to proper final spacing if direct seeded. Depending on the species[1] particular vigor, recommended spacing for perennial species ranges between 1 foot x 1 foot and 2 foot x 2 foot or 1 foot between plants and 2 to 3 feet between rows. Woody plants should be placed on 3-foot centers with more aggressive or larger species on 5-foot centers. Maintain moist conditions until plants are well established. Division is beneficial (or imperative) for many perennial species after the second year of production in order to maintain productivity.

Tall or relatively top-heavy species will require stem support. Rig beds with adjustable wire or plastic netting that can be raised as plants mature. Be sure the netting and supports are in place before the plants get too tall. It's extremely difficult to "retrofit" support without damaging the plants. Many species, such as Phlox paniculata, benefit from pinching to encourage branching and obtain the maximum number of stems per plant. Pinch as soon as the plants are well-established and elongating. Leaving some of the crop unpinched can result in earlier flowering and larger flowers with the trade-off of fewer stems. Consult crop-specific references for timing.

Use of Fertilisers

Fertiliser requirements differ from crop to crop. Some annuals, such as the annual sunflower (Helianthus annuus), are heavy feeders and require periodic fertilisation throughout the growing season. Fertiliser delivery methods range from broadcast or side dress application of granular fertiliser to fertigation.

Incorporation of a slow-release fertiliser during Spring (not Fall) tilling will give young plants a jump-start. Periodic on-site monitoring of soil and irrigation water pH and soluble salts will be a tremendous help in designing and adjusting an appropriate fertiliser programme. Limited space does not allow a full discussion on fertiliser sources and recommendations.

Pest Management

Pests for field grown cut flowers run the evolutionary gamut, from powdery mildew, aphids and Japanese beetles, to rabbits, deer, and unscrupulous passers by. Fungicides, pesticides and shotguns are all of use in the battle for maximum yield. IPM (integrated pest management) is highly recommended as a money-saving and environmentally acceptable pest and pathogen control method for field cut flower production. If you are using only organic means or biological controls to produce your crop, tell your customers.

HARVESTING

In warm weather, it is imperative that field-grown cut flowers be harvested early in the morning. Harvest when plant water status is high and temperatures and transpiration are relatively low. Wait until dew or other moisture has evaporated, if possible. Wet flowers and foliage are more susceptible to postharvest pathogens. Do not harvest when light level and temperature are at a maximum. Shading the freshly harvested material also helps maintain lower temperatures. Harvesting at the proper stage of development for each species is very important. Too early, and some species may not open; too late can result in drastically reduced vase life. Harvest is the most labor-intensive aspect of cut flower production. Communication with your harvest workers is essential! Be clear about what is acceptable and what isn't to insure uniformity of the product.

Field and handling sanitation is just as important as it is in the greenhouse business. Keep fertiliser injector systems, harvest knives or shears, postharvest handling buckets, surfaces and coolers clean and sanitised. Do your cutting, grading and bunching in one "fell swoop" to eliminate excess handling which can increase the cut's exposure to pathogens and water stress. There are numerous ways to preserve and make use of surplus cuts - air drying, oven drying, silica gel, glycerin, etc.

Postharvest Care

Proper postharvest care of your cuts is essential for maintaining high quality and a long vase life. The plant's life processes continue even after the stem is cut; respiration, transpiration, growth and development still happen. The cut stems and flowers remain sensitive to damage and disease. Floral preservatives and other additives are a necessary part of the postharvest process. Cool water can serve to promote cooling of the stems. Warm water is useful if the cuts are under extreme water stress. Monitor water pH - acidic water (pH 3.0 to 5.5) inhibits bacterial growth helping flowers persist longer. Preservatives are also formulated to be effective at lower pH. Mixing your own postharvest preservative concoctions is not recommended. There are many commercial sources for flower preservatives, conditioners, hydrators, and ethylene inhibitors. Ethylene is another important consideration affecting postharvest longevity. Flowers cannot be stored with any kind of fruit or vegetable. The ethylene produced by the fruit or veggie will result in premature floral senescence. Good ventilation and removal of dead and dying flowers are essential to maintain a relatively ethylene-free environment.

Grading, Packing and Delivery

There is no mandatory grading system for specialty cut

flowers. General rules of thumb apply, however, emphasising uniformity: no greater than 10% deviation in stem length, relative uniform stem diameter, flowers of uniform size and stage of development. Ten stems per bunch is the standard for most species, with some species sold in fives or as singles. There are myriad packing options - buckets, boxes, flowers held wet or dry; find out which are appropriate for the species you are growing. The majority of specialty cut flower growers use the indispensable 5-gallon plastic bucket. Be aware that some species such as snapdragon and gladiolus exhibit a geotropic response: stems laid flat will bend upwards, away from the gravitational pull, resulting in curved stems.

Vehicular and personnel requirements necessary for timely deliveries are often overlooked in the planning of a cut flower business. Things can get complicated, reducing efficiency and profits. One alternative is to deliver directly from the field in the morning. This works adequately for immediate delivery to local markets. However, if you need to hold the flowers for any reason, such as accumulating certain cuts for a larger wholesale market, cold storage facilities will be necessary. There are many options available at a wide range of costs - built-in-place coolers, prefabricated cold storage units, or even modified refrigerated transportation units such as refrigerated semi-tractor trailers or ice cream trucks. If using an independent shipper, be sure your carrier is educated to the need to maintain temperatures between 35° and 40° F during transit.

PLANT PROPAGATION

Most plants reproduce more of their kind through production of seeds. This is sexual propagation and it involves the exchange of genetic material between two parent plants. Many ornamental plants do not come true from seed. To increase the numbers of these plants, gardeners and horticulturists use asexual propagation. In asexual propagation, the new plants are genetically exact copies or clones of a single parent plant. The methods used in asexual propagation range from taking leaf cuttings of African violets to grafting apple cuttings onto root stocks.

Sexual Propagation

Sexual propagation of plants involves the exchange of genetic material between parents to produce a new generation. Seeds and spores can be used for repro-duction. Seeds are typically produced from sexual reproduction within a species, since genetic recombi-nation has occurred plants grown from seed may have different characteristics to its parents. Some species produce seed that requires special conditions to germinate like cold treatment. The seed of many Australian plants and plants from southern Africa and the American west require smoke or fire to germinate. Some plant species, including many trees do not produce seed until they reach maturity, which may take many years. Seed can be difficult to acquire and some plants do not produce seed at all.Sexual propagation offers the following advantages:

— It is usually the only method of producing new varieties or cultivars.

— It is often the cheapest and easiest method of producing large numbers of plants.

— It can be a way to avoid certain diseases.

— It may be the only way to propagate some species.

Propagation by Seed

Propagation by seed is the most commonly known method of producing new plants. Common annual and biennial vegetables and flowers are grown easily from seed. Perenniais grown from seed may take more than one season to flower. Lawn grasses are commonly grown from seed by the home gardener. Use seed that is guaranteed reliable for freshness and purity. Purchase seed that is packaged for the current year. Most seed companies provide information on expected percent of germination. This figure tells what percent of seed is expected to sprout or germinate. Generally, 65-80% of most fresh seed sown will germinate. Of those seeds that germinate, 60-75% will grow into satisfactory seedlings. When choosing seed, be sure to select varieties or cultivars that meet your requirements for size, colour and growth habit. Choose varieties that will mature before frost. This can be estimated by using the number of days to maturity and comparing it to the days between probable planting date and the average frost date of the area. When considering seed not packaged for the current year, remember that the germination rate of most seeds drops dramatically after the first year.

Seed catalogs are not only helpful in purchasing seed, but they are also excellent references for cultural information. Catalogs and seed packets may provide information on country of origin, bloom time, specific germination requirements, cultural requirements and disease resistance. Some seed packets indicate if seeds have been chemically treated to prevent disease. Read packets and catalogs thoroughly and follow their recommendations. The more information the gardener possesses, the greater the chance of growing plants successfully.

Seed saved by the home gardener will probably be the result of random pollination by insects or other natural mechanisms. Random pollination results in seeds that produce plants that may not be identical to the parent plant. The seeds of hybrid cultivars should not be saved. Some plants make excellent candidates for seed saving. Common self-pollinated, non-hybrid and purebred annual vegetable seeds that can be saved include lettuce, beans, peas, herbs and heirloom tomatoes. Saving seed saves money. It allows the gardener to maintain varieties that are not sold commercially. Many avid seed savers belong to groups that exchange seed through networks. Some seed saver groups specialise in keeping heirloom varieties. Many heirloom varieties are the great-grandparent plants of modern cultivars. It may be tempting to bring home seeds or plants seen on vacation in foreign countries. However, this is how many serious insect and disease pests are introduced. A nonnative plant may become a noxious weed. Follow all import regulations for horticultural materials.

It is important to save seed from healthy plants because some diseases can be carried in seeds. Commercially grown seed is protected from disease problems because it is produced under very strict conditions with frequent inspection. Harvest seed just before fruit is fully ripe. For flowers with exposed seeds, place the seed stalk or flower head in a bag and store in a warm, dry location. Seed will fall into the bag when it is completely dry. The seed of pulpy fruits should be separated from the pulp, washed and thoroughly dried.

Once seeds are completely dry, place them in airtight storage containers marked with name and date saved. Store seeds at 40 degrees F with low humidity. The refrigerator provides these conditions. Seed of many plants can remain viable for up to 5 years if properly stored. However, it is best to use home-harvested seed during the following growing season. Some species of plants produce seeds that are short-lived. These seeds must germinate immediately after they ripen or they lose their viability. Delphinium, onion and parsley are examples. Before planting, it is a good idea to check stored seed for its germination rate. Planting these seeds directly in the garden may be a waste of time and effort if germination rate is very low. To check germination rate, place some of the seeds between paper towels that are kept constantly moist and between 65 and 70 degrees F. Check the seeds daily for germination. If the germination rate is 70% or less, consider buying new seed.

Germinating seed

When germinating seed it is helpful to remember that a seed is made up of three parts:

— an outer protective coat

— a food supply under the seed coat

— an embryo of a young plant

The protective coat prevents sprouting until ideal growing conditions exist. Bringing seeds out of dormancy involves manipulating conditions to hasten germination. Even with ideal conditions, some seeds are still very difficult to germinate. There are several factors that affect germination. Water (moisture), light (or dark), oxygen and heat play a part in triggering germination. In addition to environmental factors, seed must be viable.

Water is essential in the first phase of germination. Water penetrates the seed coat and causes the endosperm to swell. The seed coat, softened by water, splits open as the endosperm swells. The water dissolves nutrients in the endosperm making them available to the embryo and growth begins.The growing medium must be constantly moist, but not wet. Any dry period may cause death of the sprouting embryo.

Light can stimulate or inhibit a seed's germination. This determines whether the seed should be sown on the surface of the growing medium or below the surface. Check the seed packet or catalog for light requirements. Oxygen is required by the embryo to begin growing. The seed must respire to break down the food stored in the seed. This is one reason for using a light, well-aerated growing medium for starting seeds.

Every seed has an optimum temperature range for germination. Many seeds have a fairly wide temperature range for germination, but some are limited to a narrow range. The temperature range is usually given on the seed packet or in the catalog. The temperatures required by many seeds are higher than those in most homes. The desired constant temperature can be achieved through heating cables placed under germination containers. Setting flats or pots on radiators, the furnace or on the refrigerator will provide bottom heat. However, these locations may be too hot and cause the soil to dry too quickly.

Seed scarification

The coat of certain seed is extremely tough and must be penetrated by special means. Particularly hard seed may be scarified. Scarification involves breaking, scratching or softening the seed coat to allow moisture penetration. Two methods of scarification commonly used by the home gardener are mechanical and hot water. Mechanical scarification involves breaking or weakening the seed coat with a file, sandpaper or hammer. Hot water scarification involves placing seeds in water that is 170 to 210 degrees F. After the water cools, seeds should continue to soak for 12 to 24 hours. Then they are planted. Specific instructions for scarification are usually mentioned on the seed packet or in the seed catalog.

Seed stratification

Some seeds will not break their dormancy unless exposed to a period of low temperature and moist conditions. Stratification requirements are usually indicated by the seed supplier. Plants that typically require seed stratification include many trees and shrubs, and certain perennials. This period of "chilling" or stratification can be accomplished by placing seed in a moist, sterile (pasteurised) growing medium, such as a mix of equal parts clean sand and peat or sphagnum peat moss, in a disinfested container. Enclose the container in a tightly sealed plastic bag and place it in the non-freezer section of the refrigerator. Towards the end of the stratification period seeds may start to sprout. Handle sprouted seeds carefully when transplanting into pots. Some seeds may require both scarification and stratification to germinate reliably.

Sowing seed Indoors

Sowing seed indoors is the easiest and cheapest method of producing vegetables, annual flowers and some perennial plants. Plants with extremely small seeds or those that need a long growing season make excellent candidates for starting indoors. Very small seed is difficult to plant outdoors because of size. Plants that require a long growing season may not have enough time to reach maturity unless started indoors in winter or early spring.

Supplies needed for indoor seed sowing include the following: fluorescent or grow lights, disinfested containers with excellent drainage, pasteurised (sterile) seed-starting medium and a location with proper temperature and ventilation.

Growing media. Choose a medium with a loose, uniform, fine texture. A pasteurised mixture that is 1/3 soil, 1/3 sand, vermiculite or perlite, and 1/3 peat moss has the qualities of a good seed-starting medium. Retail garden centers carry mixes labeled for seed starting. Whatever is selected, be sure it is pasteurised (sterile). Using pasteurised soil prevents damping-off, a fungal disease that kills young seedlings. Pasteurised soil also helps to avoid weeds, diseases and pests. Seed-starting media are usually low in fertility. This means that a regular fertilisation programme is very important once seedlings emerge.

Containers. Any recycled containers are adequate for seed starting provided they are disinfested, have good drainage and are at least 2 inches deep. Other container options include compressed peat pellets, peat pots, paper pots, plastic cell packs and flats. Peat and clay containers tend to dry more quickly than plastic containers because they are very porous.

Seed sowing. The correct timing of seed sowing is an important factor in successful indoor seed starting. In winter months, overanxious gardeners may sow seeds too soon. Seedlings that are held indoors too long perform poorly once transplanted into the garden. Most seeds should be sown 4 to 12 weeks prior to transplanting into the garden. The time it takes for seedlings to be ready for transplanting outdoors will vary.

An acclaimation period before placing seedlings directly into the permanent growing site must be included. However, readiness for outdoor planting will vary with how quickly germination occurs, the growth rate and weather conditions. Quality and quantity of light, temperature and nutrients affect the growth rate. Seed catalogs and packets provide information on days to germination and weeks needed to reach transplant size.

Fill the container to within 1/4 inch of the top of the container with moistened seed-starting medium. After sowing seeds, keep the medium moist, not wet. To keep the medium moist, you may place the container in a plastic bag just large enough for the container. Seal the bag. The plastic bag keeps moisture in, but allows air exchange. The plastic bag method should not require any further watering until germination. Provide proper light and temperature conditions. Once seedlings germinate, remove the container from the plastic bag. Place the container in a location that has high light intensity and cooler temperatures.

Use the following rules to sow seed, depending upon seed size. Sow very small seeds by sprinkling on top of the medium and pressing in. Use a fine mist of water to gently wash seed into the growing medium. Sow medium-size and larger seed in rows 1 to 2 inches apart, and 1/8 to 1/4 inches deep. If no depth is specified on the seed packet, use the general rule of planting the seed at a depth twice the diameter of the seed.

When sowing in a tray or flat, sowing in rows is preferred over scattering seed. This method provides better air circulation than scattering of seed. When seedlings are crowded, they may become tall and spindly. To avoid the need to transplant seedlings from a seed flat to pots, you may sow seeds directly into cell packs or peat pots. Plant two or three seeds per cell or pot. When they germinate, remove the two less vigorous seedlings.

Transplanting. Once the seeds have germinated, you will see two seed leaves or cotyledons. Eventually, these seed leaves will wither. Next, true leaves will form. These true leave look like the plant's typical leaves. Transplant the seedlings to individual containers once the first set of true leaves appear.Use a pencil, small stick or other narrow tool to lift seedlings from the seed flat. Plant each seedling in its own small pot filled with pasteurised growing medium. The new container should be no larger than a 2-1/2- to 3-inch pot.

Hold the seedlings by a leaf between the thumb and forefinger. Handling by the leaf avoids damaging the fragile stem which would kill the seedling. If a small leaf is lost in transplanting, the plant can recover and grow normally. Expose seedlings' roots to air as briefly as possible. Lightly firm the soil around the seedling.

Transplanting temporarily slows or stops the growth of seedlings. Sowing into cell packs bypasses the need to transplant the seedlings.

Watering. Watering can be a cause of seedling failure. Keep soil moist but not wet. Small, tender seedlings dry out rapidly and can die. Remember that roots always must have oxygen, as well as water; therefore, do not keep the soil soaked. Water when the surface of the soil begins to dry out.

Bottom watering helps prevent damage to the seedlings caused by a hard stream of water. Bottom watering also encourages deep root development and ensures that the entire depth of soil receives moisture. Do not let the pot or flat sit in water longer than it takes for all of the soil to become moist.

Temperature for seedling indoors. Keep seedlings in a well-ventilated, cool location. The temperatures should be 55 to 60 degrees F at night and 65 to 70 degrees F during the day. These temperatures encourage compact, bushy, vigorous growth while minimising disease.

Light requirements. Seedlings require bright light immediately after germination. One warm-white, 40-watt bulb and one cool-white, 40-watt bulb used together are adequate for seed starting and seedling growth. Fluorescent lights can be used for one year before replacement is recommended.

Special grow lights are also suitable, but more expensive. The lights should be no more than 6 inches above the top of the seedlings. Mount the light fixture so it can be raised as the plants grow in height.

Day-length requirements vary with different plants. Petunia, snapdragon, phlox and China aster require short daylight periods of 10 to 12 hours. However, most plants that are started from seed benefit from 16 to 18 hours of light.

Ferlilisation of Indoor Seedlings. Because the growing medium used to start seedlings is usually low in nutrients, a regular fertilisation programme is important for proper plant growth. Apply a liquid fertiliser high in phosphorous weekly. Fertiliser with a 1-2-1, N-P-K ratio is recommended. Dilute fertiliser 1/4 to 1/2 the label's recommended strength and apply sparingly. Always use a liquid form of fertiliser.

Pinching of seedlings. Pinching the growing tips of seedlings will result in more branching. This produces a fuller, stockier plant.

Hardening off Seedlings before Transplaning. Plants grown indoors must be gradually introduced to outdoor conditions. If seedlings are not hardened off, leaves may be burned by the intensity of the sun or damaged by wind. Acclimate plants by first placing them in a cool, protected location, such as a porch or shaded COLD FRAME. This first step in hardening off allows plants to adjust to outdoor temperatures. After 7 to 10 days, move seedlings into a shaded area of the garden for 2 to 3 days. This will prevent sunscald. Finally, hardened seedlings can be planted directly into the garden as weather permits. Planting on a cloudy day or late in the evening is a sensible precaution.

Transplaning Seelding into the Garden. The garden soil should be adequately dry to prevent compaction. At this point, plants will again experience transplant shock and a setback in growth. Plants must adjust to dramatically different nutrient levels, soil temperatures, moisture levels and soil tilth in the garden.

Pull apart the lower portion of the root mass to get the roots growing outward. If seedlings have been grown in peat pots, pull apart the bottom of the pot and roots. Although seedlings may be planted without removing the pot, be sure to maintain the same soil level. Trim away any of the pot that is above the soil line. The exposed portion of the peat pot acts as a wick and dries out the entire pot and roots. Water seedlings into the soil. A cup of transplanting solution will help plants get off to a good start. Make your own transplanting solution by mixing 1 tablespoons of a water-soluble 20-20-20 fertiliser in a gallon of water.

Sowing Seed Directly into the garden

Many flowers and vegetables may be sown directly into the garden. Direct sowing avoids transplant shock. It takes less work but involves more risk from weather, pests, diseases and erosion.Before sowing seeds directly into the garden, know what conditions are required for germination and growth. A warm-season crop, such as beans, may rot before germinating if planted in cold, damp soil. Knowing the average frost date for your area helps to avoid losing frost-sensitive plants. Some cool- season crops, such as peas and lettuce, should be planted early in the season while temperatures are low.

Sow seeds in a row or broadcast them into a well-raked seedbed. The seedbed should be free of stones or other large debris. Choose a calm day. To broadcast seeds, merely scatter them over a large area in the seedbed. Cover the seeds with a fine layer of soil. To sow very small seeds, mix them with sand before scattering. Then water with a gentle spray. Avoid washing seed away when watering.

Propagation by Spores

Ferns can be propagated from spores which develop in clusters on the underside of fronds. Collect fronds that have produced spores and store them in an envelope until dry. After drying, separate the dust-like spores from the cases by screening. Store spores in an airtight container in a cool, dry place until ready to plant. Germinating spores requires more time and care than germinating seeds. Growing ferns from spores involves the two different generations of ferns. Spores first produce an asexual plant called a gametophyte. This plant is very small and has none of the usual plant parts. It resembles a moss-like growth and is about 1/8 inch thick.

The gametophyte reproduces sexually and forms sprophytes which have visible roots, stems and leaves. During the first phase of growing ferns, sterile conditions are critical. Moss, fungi and algae compete aggressively with young fern gametophytes. Sow spores on top of a pasteurised (sterile), moist, soilless mix or sphagnum peat in a disinfested container. Water must also be sterile. It takes from 3 to 6 months to grow ferns from spores.

Asexual Propagation

Asexual propagation does not involve exchange of genetic material, so it almost always produces plants that are identical to a single parent. Delicious apple and Bartlett pear are two examples of species that have been asexually propagated for decades. Asexual propagation methods include cuttings, layering, division, grafting, budding and tissue culture.

Cuttings

Cuttings involve removing a piece from the parent plant and that piece then regrows the lost parts or tissues. Both woody and herbaceous plants are asexually propagated by cuttings of stems, leaves and roots. New plants can be grown from parts of plants because each living plant cell contains the ability to duplicate all plant parts and functions. Mature cells can change into Meristematic cells that are found at rapid growth sites like buds. There are many types of cuttings. Often, a plant can be propagated by more than one method of cutting. Some plants will reproduce readily from cuttings and others take a considerable amount of time and care.

Stock Plants are the parent plants used in asexual propagation. Stock plants must be in excellent health and should possess characteristics desirable for production of new plants. Herbaceous cuttings are those taken from nonwoody plants, such as perennials and houseplants.

Softwood cuttings are pieces of new growth taken from woody stock plants. These cuttings must be taken before the new growth starts to harden. Hardwood cuttings are taken from tissue which has become woody. Other forms of cuttings are leaf cuttings and root cuttings.

The gardener must try to duplicate the conditions needed for a plant to root from a cutting. High humidity, indirect light and soil temperatures of 70 to 80 degrees F are best for most cuttings. These conditions may be created by keeping cuttings enclosed under glass or in plastic bags in dappled shade. Cuttings must be shielded from direct sunlight, especially if they are under glass or plastic.

Stem cuttings

When a cutting is made, injured xylem and phloem cells plug the tubes so that precious fluids are not lost. Usually a Callus forms at the cut. Cells near the callus area reorganise to form adventitious roots. Stem cuttings are the most commonly used method to produce houseplants. Select vigorous, new growth with no flower buds. Stem sections should be free of diseases and insects. Each cutting should be 2 to 4 inches long and have 2 or 3 leaves attached.

Make a cut 1/4 inch below a leaf node and pull off the leaves that are at the nodes that will be below the surface of the rooting medium. Rooting Hormone helps to stimulate rooting, but is optional. Pour a small amount of the rooting hormone into a clean container to prevent contamination of all of your rooting hormone. Dip the base of the stem, including the node area, into the rooting powder. The stem should be dry when dipped.

Commercial rooting products often include a fungicide. This is a good idea given the damp conditions required for rooting success. Tap off excess powder, since too much hormone can inhibit rooting.

Poke a hole in the medium before inserting the cutting to avoid loss of the rooting hormone. Insert treated cutting in a moist rooting medium. A suitable rooting medium is half perlite and half sphagnum peat moss. Any disinfested container with drainage is acceptable for use.

Cover container and cutting with a plastic bag tent to maintain high humidity. Place unit in a warm area with indirect light. Check the rooting medium every few days to make sure it remains moist. Rooting can take from a few days up to several months. After a few weeks, test for rooting by gently tugging at the cutting. If there is resistance, rooting has started and the plastic cover may be removed.

Leaf cuttings

In this method, a leaf blade or leaf with petiole is used to propagate new plants. The same steps are followed as for stem cuttings. Choose a healthy leaf from a vigorously growing plant. Cut it close to the stem with a sharp, disinfested razor or knife. Trim off 1/4 of the leaf and dip into rooting hormone, if desired. Insert the leaf into rooting medium so that 1/3 of the leaf is below the surface.

One or many new small plants form at the base of the leaf. African violet leaves will produce many new plants. Begonia leaves can be divided into segments for propagation; however each leaf piece must contain a major vein. With leaf cuttings, the original leaf is not a part of the new plant and is usually discarded. Many succulent plants, such as sedum, jade, and peperomia, can be propagated by leaf cutting. Some plants such as Kalanchoe pinnata or K. bryophyllum, piggy-back or air plants, form their own new plantlets in this fashion.

Root cuttings

Cultivation of root cuttings probably started after gardeners observed new plants growing from pieces of root accidentally left behind in the soil. Take cuttings from newer root growth. Make cuttings 1 to 4 inches long from roots that are 1/4 to 1/2 inch in diameter. Be sure that the roots collected are from the chosen plant and not neighboring plants. Cuttings should be taken during the dormant season when roots have large carbohydrate supplies. However, they also may be taken throughout the growing season. Cut straight through the end of the root closest to the stem. Cut the other end on a slant. This allows you to remember which end is the top (the straight cut) and which is the bottom (the diagonal cut).

Store cuttings from dormant roots for 3 weeks in moist rooting medium at 40 degrees F. Remove from storage and plant upright in the growing medium. Keep moist and warm, in a bright location until growth and weather permit acclimatising to the outdoors. If root cuttings are taken during active growth, skip the storage period and place cuttings directly in the rooting medium. For smaller plants, take 1- to 2-inch sections. Place cuttings horizontally a half inch below the surface of the rooting medium. These cuttings should be handled indoors or in a HOTBED. The fine roots of many perennials are used for propagation. Root cuttings of some variegated plants will lose their variegation.

Softwood and hardwood cuttings

Softwood cuttings are taken from first-year branches that have not yet become woody. Flowering shrubs are often propagated by softwood cuttings. Late spring and early summer are the best times for success with this method. Take cuttings after it rains or water is applied in the cool morning. Make a diagonal cut. The larger diagonal cut gives more area to develop roots. Keep cuttings in water before moving them into rooting medium.

Make cuttings 2 to 10 inches long. Larger cuttings produce larger plants sooner. Make cuts slightly below a leaf node. Follow the same instructions for disinfested containers, rooting medium and light and moisture conditions used by other cutting methods. Rooting hormones may be used with this method. The base of the cutting should be dry before dipping it into rooting hormone powder.

Hardwood cuttings are taken once the tissue becomes woody and the plant is dormant. Cuttings can be taken anytime from late fall after a killing frost until late winter. Select healthy wood that was produced the previous summer. Several cuttings can be made from the same branch of some shrubs.

Make cuts at a slant, 5 to 12 inches long. Basal cuts should be just below a node, while the upper cut should be slightly above a bud. Dip the basal end in rooting hormone. Mark the most terminal end of each cutting with a tag. Bury cuttings vertically in moist vermiculite or sand. Cuttings should not freese, but must remain cool. A callus will form on the lower cut end during storage. Callus formation indicates that cuttings are ready to root.

Layering

Layering causes roots to develop on shoots that are still attached to the parent plant. The stem is not cut from the main plant until it has rooted. Use this method when other propagation methods are unsuccessful. Layering is a good propagation choice when only a few plants are needed.

Simple layering is done by bending a branch to the ground and burying a portion of it. The tip remains uncovered. A light soil increases rooting success, as will wounding or girdling the buried section. Treatment with rooting hormone is helpful. Hold the branch in place with a rock or peg. Layering is done in early spring while plants are still dormant or in late summer on wood that has not become woody. Plants with flexible branches are particularly suited to this method. Other types of layering include serpentine or compound layering, continuous or trench layering, and mound or stool layering.

Divisions

Division is the cutting or breaking up of a crown or clump of suckers into segments. Each segment must have a bud and some roots. These segments are replanted and grow into new plants identical to the parent. Most perennials should be lifted and divided when they become overgrown and begin to lose vigor. Vigorous growth in most perennials occurs on the outer segments of the clump. Old growth in the center of the clump is discarded.

Carefully dig the plant, loosening the roots and lifting the plant from the soil. Split apart the main clump with two spades or forks or chop with a shovel or hatchet if the clump is firmly massed. In some cases outside segments of the plant can be removed and replanted without disturbing the rest of the plant. A good rule of thumb is to divide fall-flowering perennials in spring and spring- and summer-flowering perennials in fall. Some shrubs that form clumps or crowns are suitable for division. Divide in early spring while dormant. Shrubs used for division should have several clumps that are two years old or older.

Bulbs and corms

Bulbs can be propagated by removing small bulblets or offsets that form at the base of the parent bulb. These small bulbs take 2 or 3 years to mature into plants that flower. Place offsets in rich, light soil for their development. This same procedure should be followed for plants which form from corms, such as gladiolus.Many lilies can be multiplied by removing scales from the mature bulb. Dust the scale with a fungicide and place, base end down, in a moist growing medium in a warm, protected area. Bulblets will form at the base of the scale.

Tubers and rhizomes

Tuberous plants can be dug up and the tubers separated. In separating the tubers, each must have a segment of the crown that contains at least one eye or bud. Rhizomes grow and develop buds along their length. The rhizomes can be dug and cut into sections that each contain at least one eye or bud.

Grafting

Grafting involves the joining of different segments of two different plants of the same species. A branch or bud is joined to an understock. In grafting, the cambium layers of the two different segments are aligned and grow together. Some plants graft naturally. Ivy commonly grafts itself. Natural grafting occurs where two branches are in close contact over several years.Grafting allows gardeners to produce plants identical to a parent plant. It also allows growers to control size and shape of a tree or shrub. Apples are grafted onto root stock that regulates the ultimate size of the tree.

Grafting can produce more vigorous, earlier fruiting, or two varieties can be grown on the same tree to provide pollination as with apples. On the negative side, some grafting attempts will be rejected. This is called graft incompatibility. This might not occur in the first year, but may in successive years. Wind or winter cold can damage a graft union. Some grafted trees or plants produce large numbers of suckers. These can crowd out the desired plant or tree and are unsightly. If the graft union is planted below the soil surface, the scion will root and the rootstock will die.

Grafting is usually done in the spring. It involves collecting small branches called scion wood. These branches are 1/4 to 1/2 inch in diameter. Gather them several weeks before grafting is to occur. Collect wood that grew the preceding year. Select only wood with leaf buds, not flower buds. Scion wood should be gathered in winter when wood is dormant, but not frozen. Use sharp clippers so that wood is not crushed. New growth over 1 foot in length is usually best. Discard the wood at both ends of the branch and use the middle section. End growth is too succulent for grafting and wood closer to the base will graft more slowly. Label the scion wood, wrap it in moist paper towels or sphagnum peat, enclose it in an airtight, plastic container and place it in the refrigerator.

Scion wood must be joined to the understock in spring when buds swell. It is critical that the two pieces are nearly the same size and that sap has begun to flow. The day before actually grafting, remove scions from the refrigerator and snip off the bottom ends. Place the clipped scions in a pail of water overnight. It is critical that the cambium layer on the scion precisely matches that of the understock.

Grafts can be placed on a root or stem and be joined by any of several methods. The union is held firmly in place using wax, tape, plastic wrap, rubber bands or plastic tubing.The grafted area must be protected from anything that will move the scion out of alignment. Bracing some grafts is advised. Side sprouts should be removed the first summer. Sucker shoots that sprout from below the graft also should be removed. If growth of new graft is satisfactory, do not fertilise the plant during the first year. There is danger that top growth will be very brittle and it could fail to harden off before the first frost.

Bud grafting

Bud grafting is faster, easier and less messy than other forms of grafting. Cambium layers do not need to be aligned. Bud grafting is done from early July through early August. This method uses a newly developed latent bud, taken from under a live leaf. Budwood is collected from healthy branches that grew since spring. Young trees offer good budwood because they produce a large amounts of new growth. Use buds from the middle section of the branch.

Remove leaves, but keep 1/4 inch of the leaf's petiole for grasping. The bud is cut from the branch and inserted into a T-shaped slice made in the bark of the understock. Budding should only be done when the bark slips easily away from the tree. The bud is held in place with special tape or wrap.

Micropropagation

Micropropagation is the practice of rapidly multiplying stock plant material to produce a large number of progeny plants, using modern plant tissue culture methods. Micropropagation is used to multiply novel plants, such as those that have been genetically modified or bred through conventional plant breeding methods. It is also used to provide a sufficient number of plantlets for planting from a stock plant which does not produce seeds, or does not respond well to vegetative propagation. Each plant cell has the potential to grow into a new plant exactly like the parent. This fact coupled with technical advances, specialised equipment and sterile laboratory conditions has produced tissue culture. In tissue culture, individual or small groups of plant cells are manipulated so they each produce a new plant. A tiny piece of bud, leaf or stem can produce incredible numbers of new plants in a small space in a short time.

The advantages of tissue culture, in addition to speed and efficiency of propagation, include production of disease-free plants. New plants can be made available to the public more quickly because of tissue culture. However, there are some problems with spontaneous mutations which naturally occur. In tissue culture, the incidence of these mutations is greatly increased. Conditions for tissue culture are very exacting. Absolutely sterile conditions must be maintained. Temperature, light, humidity and atmosphere are strictly controlled with electronic sensors and computerised controls. Such costly equipment rules this out for most home gardeners.

Micropropagation begins with the collection of a sterile explant(s). This small portion of plant tissue, which may be as small as a cell, is placed on a growth medium, typically a medium containing sucrose as an energy source and one or more plant growth regulators (plant hormones).Usually the medium is thickened with Agar to create a gel which supports the explant during growth.The plant tissue should now begin to grow and differentiate into new tissues. For example, media containing cytokinin are used to create branched shoots from plant buds.

Following the successful growth of plant tissue, the establishment stage may be repeated, by taking tissue samples from the plantlets produced in the first stage. Through repeated cycles of this process, a single cell sample may be magnified to hundreds or thousands of plants.

Pretransplanting is stage involves treating the plantlets/shoots produced to encourage root growth and "hardening". It is performed in vitro, or in a sterile "test tube" environment. Root growth does not always occur in the earlier stages in plant cell culture, and is of course a requirement for successful plant growth after the micropropagation procedure. It is performed in vitro by transferring the plantlets to a growth medium containing auxin(s).

"Hardening" refers to the preparation of the plants for a natural growth environment. Until this stage, the plantlets have been grown in "ideal" conditions, designed to encourage rapid growth. Due to lack of necessity, the plants are likely to be highly susceptible to disease and will be inefficient in their use of water and energy. Hardening typically involves slowly weaning the plantlets from a high-humidity, low light, warm environment to what would be considered a normal growth environment for the species in question. This stage (pretransplant) is not always performed, instead being incorporated into the last stage by encouraging root growth and hardening ex vitro, or in nonsterile plant media. In the final stage of plant micropropagation, the plantlets are removed from the plant media and transferred to soil or (more commonly) potting compost for continued growth by conventional methods. This stage is often combined with the "Pretransplant" stage.

Micropropagation has a number of advantages over traditional plant propagation techniques:

— Micropropagation produces disease-free plants
— Micropropagation produces rooted plantlets ready for growth, rather than seeds or cuttings
— It has an extraordinarily high fecundity, producing thousands of propagules in the same time it would take a conventional technique to produce tens or hundreds
— It is the only viable method of regenerating genetically modified cells or cells after protoplast fusion
— It is a good way of multiplying plants which produce seeds in uneconomical amounts (if at all)
— Micropropagation often produces more robust plants, leading to accelerated growth compared to similar plants produced by conventional methods

IMPROVING FLOWER PRODUCTION

The availability of natural resources like favourable and diverse climatic conditions permit production and availability of a large variety of flower crops round the year. Cheap labour leads to reduction in production costs, increasing access of the consumer to good quality flowers at affordable prices, besides increasing our competitiveness in the export markets. Being a new concept in the agri-business, it took some time for scientific commercial flower production to take roots, but with the appreciation of its potential as an economically viable diversification option, its growth is slowly stabilising. The government also has, during the last few years, recognised floriculture as an important segment for developmental initiatives.

Model Floriculture Centres being set up in 11 major production zones, to serve as focal units for development in the region, have a mandate of making available quality planting material, new/improved production technologies and also to provide training in production and post harvest management. There are also special government programmes for area expansion in floriculture with state assistance. The National Horticulture Board, a major developmental agency for horticulture, also makes available finances as soft loan for setting up integrated projects for production and marketing. As mentioned earlier, the government is investing in improving the infrastructure for marketing in the domestic sector.

Production of cut flowers for exports is also a thrust area for support. The Agricultural and Processed Food Products Export Development Authority (APEDA), the nodal organisation for promotion of agri-exports including flowers, has introduced several schemes for promoting floriculture exports from the country. These relate to development of infrastructure, packaging, market development, air freight subsidy etc. The 100% Export Oriented Units are also given benefits like duty free imports of capital goods. All these efforts indicate the government's commitment for improving the sector and creating a positive environment for entrepreneurship development in the field.

Being a new concept, the requirements of scientific and commercial floriculture is not properly understood in the country. The developmental initiatives of the government have to keep in mind the low knowledge base, small land holdings, unorganised marketing and poor infrastructural support. While long experience of flower growing in the open field conditions enable sufficient flower production for domestic markets, the quality of the produce, in view of its exposure to various kinds of biotic and abiotic stresses, is not suitable for the ever growing export market. The production technology for flowers under protected environment of green houses needs to be standardised. There is hardly any post harvest management of flowers for the domestic market.

Availability of surplus flowers from exports for sale in the domestic market, has increased the appreciation of quality produce and the demand for good quality flowers is increasing. With the introduction of new varieties of crops in the country, facilities for generating their planting material for large scale production need strengthening. Special attention needs to be paid to strengthen the marketing infrastructure like organised marketing yards, auction platforms, controlled condition storage chambers etc. Greater research efforts are also needed for integrated pest management, development of location specific package of practices for traditional flowers, value addition to traditional flowers etc.

The initial cost and availability of finance is a critical matter in the development of large commercial projects requiring heavy investments. More options for developmental finance, such as the soft loan scheme of the National Horticulture Board need to be identified. In the initial years of commercial floriculture development, the governmental support in terms of subsidies etc. needs special attention. The potential for growth of export market is always linked to the strength of domestic market - its capacity to absorb surplus and over production, and quality consciousness of consumers. Though we have a large domestic market, the marketing system and facilities need to be modernised.

The production for exports at present has suffered due to a few constraints. While our growers have been successful in producing world class quality at low cost, high air freight rates, low cargo capacity available, imposition of import duties, inadequate export infrastructure etc. have reduced their competitiveness. There is also a shortage of trained manpower to handle commercial floriculture activity. The demands of the growing export oriented industry would require adequate attention to be paid for human resource development, particularly at the supervisory level.

MARKETING

Marketing of flowers is very unorganised at present. In most metropolitan cities, with large market potential, flowers are brought to wholesale markets, which mostly operate in open yards. A few large flower merchants generally buy most of the produce and distribute them to local retail outlets after significant mark up. The retail florist shops also usually operate in the open on-road sides, with different flowers arranged in large buckets. In the metros, however, there are some good florist show rooms, where flowers are kept in controlled temperature conditions, with considerable attention to value added service. The government is now investing in setting up of auction platforms, as well as organised florist shops with better storage facilities to prolong shelf life.

The packaging and transportation of flowers from the production centres to the wholesale markets at present is very unscientific. The flowers, depending on the kind, are packed in old gunny bags, bamboo baskets, simple cartons or just wrapped in old newspapers and transported to markets by road, rail or by air. The mode of transportation depends on the distance to the markets and the volume. Mostly, flowers are harvested in the evening time and transported to nearby cities by overnight trains or buses.

In recent years, the government has provided some assistance for buying refrigerated carriage vans. A large number of export oriented units have built up excellent facilities of pre-cooling chambers, cold stores and reefer vans and their produce coming for domestic market sales are thus of very good quality and have longer vase life and command higher price. The government programmes for floriculture development include creating common facilities of cool chain in large production areas to be shared on cooperative basis. Formation of growers' cooperatives/associations are being encouraged.

In view of the unorganised set up, it is difficult to estimate the size of flower trade, both in terms of volume and value. A study conducted in 1989 estimated the trade to be worth Rs. 2050 million. It is in the period of the last five years or so that this business has really boomed in India, which is reflected in the number of new florist outlets in all cities and increase in the public's purchase of flowers as gifts. This would put the current trade at several times the earlier estimate. A recent study of Delhi market alone put the value of flowers traded on wholesale as Rs. 500 million.

The value of cut flower export from India has increased twenty five fold during the last five years. With more export oriented units coming into operation, exports are likely to grow further in the coming years. The major share of the export trade is for roses, in addition to orchids, gladiolus etc. The major markets are Europe and Japan. The exports of roses to Japan, have really picked up in the three years from Rs. 360 million in 1993-94 to Rs. 6090 million in 1995-96. As per the estimates for 1996-97, India has been the largest supplier of roses to Japan (volume wise).

The internation cut flower market has changed dramatically over the years. The Dutch dominated the flower growing market in the '80s with new varieties. Competition from Third World countries increased as they took a lion's share of the "traditional" domestic cut flower market. Rising freight costs put a damper on shipping from one coast to the other. The cut flower market is based on supply and demand, but you can often create your own demand by offering high quality and unusual product with reliable service to back it up. Crops produced in the cooler "shoulder" seasons can demand a higher price since the market is not typically flooded with product. Especially in reference to the small or starting grower, the following phrase is repeated over and over in both the research and industry literature: quality sells. Grow for quality, and don't be afraid to charge for it - price will follow quality. The profitable grower does not wait until he or she has harvest-ready flowers to decide their fate. Have a clear market plan established ahead of time. Your target market influences what you will grow, how it will be handled and packaged, and most important, your capital investment.

Cut flowers usually are sold by the bunch, in arranged bouquets, or individually. Marketing options include selling to wholesalers, florists or other retail outlets, or direct to the public. The obvious route when first starting out is to target local markets. As your volume increases, you may want to deal with wholesalers and distributors. "Direct to the public" sales include farmer's markets, roadside stands, or cut-it-yourself arrangements. How much you wish to interact with the public can help determine if a cut-your-own business is right for you. On-site sale, whether in the field or from a stand, requires a good location in a populated area, and a desire to deal with (not a mere tolerance of) the public. Production considerations may be altered a bit for the "cut your own" concept. If you are integrating a field cut flower business with a preexisting greenhouse or nursery that retails to the public, the additional marketing requirements would be minimal.

There are many direct-sales opportunities away from the farm as well, though transportation now becomes a cost factor. Farmer's markets offer a low-overhead market for novice growers. You can experiment with displays, mixed bouquets, dried materials, etc., and enjoy relatively autonomous pricing. There are many other retail opportunities available, such as restaurants, banks, etc. Craft shows are great sales venues for preserved and dried plant material.

REFERENCES

Clark D.G., B. Underwood, K. Shibuya, H. Loucas, D. Tieman, and H. Klee, "Microarray Analysis of Floral Senescence in Petunia",

Proceedings of the NATO Advanced Research Workshop on Biology and Biotechnology of the Plant Hormone Ethylene III, M. Vendrell, H. Klee, J.C. Pech, F. Romojaro (eds.). IOS Press, 2003.

Clark, D., H. Klee and A Dandekar, "Despite benefits, commercialization of transgenic horticultural crops lags", *California Agriculture*, 2004.

Dole, J.M. and H.F. Wilkins, *Floriculture principles and species*, 2 nd Edition. Pearson Education, Inc., Upper Saddle River, N.J. Available in Branson Library, 2005.

Larson, R.A., *Introduction to floriculture*, Academic Press, N.Y. [Branson Library], 1992.

Seeley, J.G., *Advancemcnts in Commercial Floriculture*, Southern Florist and Nurseryman, 1982.

8

Horticultural Harvesting

Harvesting is the gathering of plant parts that are of commercial interest. These include: Fruits – e.g. tomatoes, peppers, apples, kiwifruits, etc.; root crops – e.g. beets, carrots etc; Leafy vegetables–spinach and Swiss chard; Bulbs–onions or garlic; Tubers–potatoes; stems–asparagus; petioles–celery and Inflorescences–broccoli, cauliflower etc. Harvest marks the end of the growing period and the commencement of market preparation or conditioning for fresh products.

Harvesting can be performed by hand or mechanically. However, for some crops – e.g. onions, potatoes, carrots and others–it is possible to use a combination of both systems. In such cases, the mechanical loosening of soil facilitates hand harvesting. The choice of one or other harvest system depends on the type of crop, destination and acreage to be harvested. Fruits and vegetables for the fresh market are hand harvested while vegetables for processing or other crops grown on a large scale are mainly harvested mechanically. The main advantages of mechanised harvesting are speed and the reduced costs per ton harvested. However, because of the risk of mechanical damage, it can only be used on crops that require a single harvest.

A decision to purchase equipment requires careful evaluation of: the initial investment required, maintenance costs and the long period in which equipment may have to stand idle. In addition to this, the entire operation needs to be designed specifically for mechanised harvesting–distances between rows, field levelling, pesticide spraying, cultural practices to varieties which can be adapted to rough handling. Market preparation and the trade should also be able to handle large volumes of produce. Hand harvesting is particularly suitable for crops with an extended harvest period. The rate of harvesting can be increased by hiring more workers if, for example, due to climate, there is accelerated ripening and a need to harvest the crop quickly.

The main benefit of hand harvesting over mechanised harvesting is that humans are able to select the produce at its correct stage of ripening and handle it carefully. The result is a higher quality product with minimum damage. This is important for tender crops. However, adequate training, including supervision of the harvest crew, is required. Contractual arrangements with harvest labour also influences the final quality of product harvested. When wages are paid per week, fortnight or month, harvesting is undertaken carefully. However, when payment is per box, meters of row or number of harvested plants, harvesting can be careless.

Establishing teams and division of labour also influences quality. Long working days and/or few breaks as well as extremely adverse conditions can result in unnecessary rough handling of produce. Harvest labour needs to be adequately trained to give them the necessary skills to select produce at the correct stage of ripeness or degree of maturity as well as sorting techniques to minimise damage.

HARVESTING VEGETABLES

One of the main point behind growing vegetables is how much better they taste fresh from the garden. The only failsafe way to guarantee harvesting vegetables at the optimal time is to taste test. However, there are some guidelines for judging when vegetables are ripe and ready for harvest. Keep in mind that great flavour isn't a simple matter of size or color. Without the right combination of soil, sun and water many vegetables can vary greatly in taste and performance.

There are no precise guidelines as to when to harvest your vegetables, but there are some rules of thumb to guide you. Most vegetables are harvested just before full maturity, for maximum flavour and the most pleasant texture. The following are vegetable harvesting criteria for judging whether your vegetables are ready for picking.

— *Asparagus*: Begin harvesting when spears are 6-8 inches tall and about as think as your small finger. Snap them off at ground level and new spears will continue to grow. Stop harvesting about 4-6 weeks after the initial harvest, to allow the plants to produce foliage and food for themselves.

— *Beans (Snap):* Pick before the you can see the seeds bulging. They should snap easily into two. Check daily. It doesn't take long for beans to go from tender to tough.

— *Beets*: You can harvest and eat the green tops that you thin out of the rows. Beets are really a matter of personal preference when it comes to the right size for harvesting. They are ready any time after you see the beets shoulders protruding at the soil line.

— *Broccoli*: We eat the unopened flower buds of broccoli, so check frequently, especially as the weather warms up, to ensure you don't let the flower heads bloom. Don't expect your home grown broccoli to get to the size of supermarket heads. Harvest when the buds are about the size of a match head.

— *Brussel Sprouts*: The sprouts will mature from the bottom up. You can begin harvesting once the sprouts are at least an inch in diameter. Harvest by twisting off or cutting the sprout from the stem.

— *Cabbage*: The cabbage head will feel solid when gently squeezed. Cabbage needs to be harvested when it reaches maturity or it will continue to grow and split open.

— *Carrots*: Carrots can be hard to judge. The tops of the carrot will show at the soil line and you can gage when the diameter looks right for your variety. If the diameter looks good, chances are the length is fine too. But you will need to pull one to be certain. Carrots can be left in the ground once mature. A light frost is said to improve and sweeten the carrot's flavour.

— *Cauliflower:* As with broccoli, your home grown cauliflower heads will probably never match supermarket size. Harvest when the head looks full and while the curds of the head are still smooth.

— *Corn*: About 3 weeks after the silks form, they will turn dry and brown. The kernels should exude a milky substance when pricked.

— *Cucumber*: Cucumbers race to the harvest with zucchini. Check daily and harvest young. Timing and length will vary with variety. The fruits should be firm smooth. Over ripe cucumbers can be very bitter or pithy, even before they start to turn yellow.

— *Eggplant*: Slightly immature fruits taste best. The fruits should be firm and shiny. Cut rather than pulling from the plant.

— *Garlic*: The garlic tops will fall over and begin to brown when the bulbs are ready. Dig, don't pull, and allow to dry before storing. It's best to simply brush off the dirt, rather than washing.

— *Kale*: Kale leaves can be throughout the season. They should be a deep green with a firm, sturdy texture. Kale flavour is best in cooler weather.

— *Kohlrabi*: For the best texture, harvest once the kohlrabi bulb has reached about 2-3 inches in diameter. The bulbs become tougher as they grow and age. Pull or slice at the base.

— *Leeks*: Harvest leeks when they are about 1 inch in diameter.

— *Lettuce (Head):* Harvest once the head feels full and firm with a gentle squeeze. Hot weather will cause it to bolt or go to seed rather than filling out.

— *Lettuce (Leaf):* Harvest the outer leaves one the plant has reached about 4 inches in height. Allow the younger, inner leaves to grow. Leaf lettuce can be harvested in this fashion for most of the summer.

— *Muskmelon*: There are many varieties of muskmelon, but a general rule of thumb is that the color should change to beige and the fruit will 'slip' from the vine when lifted. You should also be able to notice a sweet smell when ripe.

— *Onions*: Onions can be dug once the tops have ripened and fallen over. Allow the onions to dry in the sun.

— *Parsnips*: Parsnips taste best if they are left in the ground until after a frost or two. They can be left in the ground over the winter and harvested in the spring. In cold areas, they should be mulched for the winter.

— *Peas*: The pea pods should look and feel full. Peas are sweeter if harvested before fully plumped. Peas really need to be tasted to determine if they are sweet enough.

— *Potatoes*: 'New' potatoes can be harvested when the tops start to flower. Carefully dig at the outer edges of the row. For full size potatoes, wait until the tops of the potato plants dry and turn brown. Start digging from the outside perimeter and move in cautiously to avoid slicing into potatoes.

— *Pumpkins*: Once the pumpkins have turned the expected color and the vines are starting to decline, they can be cut from their vines. Don't leave them out if a hard frost is expected.

— *Radishes*: Radishes mature quickly. You will see the shoulders of the bulbs popping out of the soil line. If left too long, they will become tough and eventually go to seed.

— *Rutabaga*: The bulbs should be about 3 inches in diameter, generally about 3 months after setting out. Rutabagas can be mulched, left in the ground and dug up as needed. Cold weather improves their flavour.

— *Swiss Chard*: As with leaf lettuce. Cut the outer leaves and allow the center to continue growing.

— *Spinach*: Spinach goes to seed quickly. Harvest by cutting at the soil line before you see a flower stalk beginning to shoot up.

— *Squash (Summer):* Pick young and check often. The skins should be tender enough to poke your fingernail through.

— *Squash (Winter):* Color is a good indicator of winter squash maturity. When the squash turns the color it is supposed to be, cut from the vine. Do not let winter squash be exposed to frost.

— *Tomatoes*: Harvest tomatoes when they are fully coloured and slightly soft to the touch. Gently twist and pull from the vine.

— *Turnips*: The turnip shoulders should be about 2 to 2 ½ inches in diameter at the soil line, when ready. Harvest once they reach maturity. Overripe turnips become woody.

— *Watermelon*: The white spot on the bottom of the melon should change to a deep yellow when ripe. Some people can hear a change in the sound made when the melon is thumped with a finger. It should make a hollow sound when ripe, but this is a skill that must be developed.

DETERIORATION REACTIONS

Enzymic Changes

Enzymes which are endogenous to plant tissues can have undesirable or desirable consequences. Examples involving endogenous enzymes include a) the post-harvest senescence and spoilage of fruit and vegetables; b) oxidation of phenolic substances in plant tissues by phenolase (leading to browning); c) sugar–starch conversion in plant tissues by amylases; d) post-harvest demethylation of pectic substances in plant tissues (leading to softening of plant tissues during ripening, and firming of plant tissues during processing). The major factors useful in controlling enzyme activity are: temperature, water activity, pH, chemicals which can inhibit enzyme action, alteration of substrates, alteration of products and pre-processing control.

Chemical Changes

Deterioration in sensory quality

The two major chemical changes which occur during the processing and storage of foods and lead to a deterioration in sensory quality are lipid oxidation and non-enzymatic browning. Chemical reactions are also responsible for changes in the colour and flavour of foods during processing and storage. Lipid oxidation rate and course of reaction is influenced by light, local oxygen concentration, high temperature, the presence of catalysts and water activity. Control of these factors can significantly reduce the extent of lipid oxidation in foods.

Non-enzymic browning is one of the major causes of deterioration which occurs during storage of dried and concentrated foods. The non-enzymic browning, or Maillard reaction, can be divided into three stages:

a) early Maillard reactions which are chemically well-defined steps without browning;

b) advanced Maillard reactions which lead to the formation of volatile or soluble substances; and

c) final Maillard reactions leading to insoluble brown polymers.

Colour changes

Chlorophylls: Almost any type of food processing or storage causes some deterioration of the chlorophyll pigments. Phenophytinisation is the major change; this reaction is accelerated by heat and is acid catalysed. Other reactions are also possible. For example, dehydrated products such as green peas and beans packed in clear glass containers undergo photo-oxidation and loss of desirable colour.

Anthocyanins: These are a group of more than 150 reddish water-soluble pigments that are very widespread in the plant kingdom. The rate of anthocyanin destruction is pH dependent, being greater at higher pH values. Of interest from a packaging point of view is the ability of some anthocyanins to form complexes with metals such as Al, Fe, Cu and Sn. These complexes generally result in a change in the colour of the pigment and are therefore undesirable. Since metal packaging materials such as cans could be sources of these metals, they are usually coated with special organic linings to avoid these undesirable reactions.

Carotenoids: The carotenoids are a group of mainly lipid soluble compounds responsible for many of the yellow and red colours of plant and animal products. The main cause of carotenoid degradation in foods is oxidation. The mechanism of oxidation in processed foods is complex and depends on many factors. The pigments may auto-oxidise by reaction with atmospheric oxygen at rates dependent on light, heat and the presence of pro- and antioxidants.

Flavour changes

In fruit and vegetables, enzymically generated compounds

derived from long-chain fatty acids play an extremely important role in the formation of characteristic flavours. In addition, these types of reactions can lead to significant off-flavours. Enzyme-induced oxidative breakdown of unsaturated fatty acids occurs extensively in plant tissues and this yield characteristic aromas associated with some ripening fruits and disrupted tissues. The permeability of packaging materials is of importance in retaining desirable volatile components within packages, or in permitting undesirable components to permeate through the package from the ambient atmosphere.

Degradation of nutritional quality

The four major factors which affect nutrient degradation and can be controlled to varying extents by packaging are light, oxygen concentration, temperature and water activity. However, because of the diverse nature of the various nutrients as well as the chemical heterogeneity within each class of compounds and the complex interactions of the above variables, generalisations about nutrient degradation in foods will inevitably be broad ones.

Vitamins: Ascorbic acid is the most sensitive vitamin in foods, its stability varying markedly as a function of environmental conditions such as pH and the concentration of trace metal ions and oxygen. The nature of the packaging material can significantly affect the stability of ascorbic acid in foods. The effectiveness of the material as a barrier to moisture and oxygen as well as the chemical nature of the surface exposed to the food are important factors. For example, problems of ascorbic acid instability in aseptically packaged fruit juices have been encountered because of oxygen permeability of the package and the oxygen dependence of the ascorbic acid degradation reaction.

Also, because of the preferential oxidation of metallic tin, citrus juices packaged in cans with a tin contact surface exhibit greater stability of ascorbic acid than those in enamelled cans or glass containers. The aerobic and anaerobic degradation reactions of ascorbic acid in reduced-moisture foods have been shown to be highly sensitive to water activity, the reaction rate increasing in an exponential fashion over the water activity range of 0.1-0.8.

Physical Changes

One major undesirable physical change in food powders is the absorption of moisture as a consequence of an inadequate barrier provided by the package; this results in caking. It can occur either as a result of a poor selection of packaging material in the first place, or failure of the package integrity during storage. In general, moisture absorption is associated with increased cohesiveness. Anti-caking agents are very fine powders of an inert chemical substance that are added to powders with much larger particle size in order to inhibit caking and improve flowability.

Studies in onion powders showed that at ambient temperature, caking does not occur at water activities of less than about 0.4. At higher activities, however, (aw > 0.45) the observed time to caking is inversely proportional to water activity, and at these levels anti-caking agents are completely ineffective. It appears that while they reduce inter-particle attraction and interfere with the continuity of liquid bridges, they are unable to cover moisture sorption sites.

Biological Changes

Mcirobiological changes: Micro-organisms can make both desirable and undesirable changes to the quality of foods depending on whether or not they are introduced as an essential part of the food preservation process or arise unintentionally and subsequently grow to produce food spoilage. The two major groups of micro-organisms found in foods are bacteria and fungi, the latter consisting of yeasts and moulds. Bacteria are generally the fastest growing, so that in conditions favourable to both, bacteria will usually outgrow fungi.

Foods are frequently classified on the basis of their stability as non-perishable, semi-perishable and perishable. For example, hermetically sealed and heat processed foods are generally regarded as non-perishable. However, they may become perishable under certain circumstances when an opportunity for recontamination is afforded following processing. Such an opportunity may arise if the can seams are faulty, or if there is excessive corrosion resulting in internal gas formation and eventual bursting of the can.

Spoilage may also take place when the canned food is stored at unusually high temperatures: thermophilic spore-forming bacteria may multiply, causing undesirable changes such as flat sour spoilage. Low moisture content foods such as dried fruit and vegetables are classified as semi-perishable. Frozen foods, though basically perishable, may be classified as semi-perishable provided that they are properly stored at freezer temperatures.

The majority of foods are classified as perishable unless they have been processed in some way. Often, the only form of processing which such foods receive is to be packaged and kept under controlled temperature conditions. The species of micro-organisms which cause the spoilage of particular foods are influenced by two factors:

a) the nature of the foods and

b) their surroundings.

These factors are referred to as intrinsic and extrinsic parameters. The intrinsic parameters are an inherent part of the food: pH, aw, nutrient content, antimicrobial constituents and biological structures. The extrinsic parameters of foods are those properties of the storage environment that affect both the foods and their microorganisms. The growth rate of the micro-organisms responsible for spoilage primarily depends on these extrinsic parameters: temperature, relative humidity and gas compositions of the surrounding atmosphere. The protection of packaged food from contamination or attack by micro-organisms depends on the mechanical integrity of the package, and on the resistance of the package to penetration by micro-organisms.

Metal cans which are retorted after filling can leak during cooling, admitting any microorganisms which may be present in the cooling water, even when the double seam is of a high quality. This fact is widely known in the canning industry and is the reason for the mandatory chlorination of cannery cooling water. Extensive studies on a variety of plastic films and metal foils have shown that microorganisms cannot penetrate these materials in the absence of pinholes. In practice, however, thin sheets of packaging materials such as aluminium and plastic do contain pinholes. There are several safeguards against the passage of micro-organisms through pinholes in films:

— because of surface tension effects, micro-organisms cannot pass through very small pinholes unless the micro-organisms are suspended in solutions containing wetting agents and the pressure outside the package is greater than that within;
— materials of packaging are generally used in thicknesses such that pinholes are very infrequent and small;
— for applications in which package integrity is essential, adequate test methods are available to assure freedom from bacterial recontamination.

Macrobiological changes: Warm humid environments promote insect growth, although most insects will not breed if the temperature exceeds about 35 C° or falls below 10 C°. Also many insects cannot reproduce satisfactorily unless the moisture content of their food is greater than about 11%. The main categories of foods subject to pest attack are cereal grains and products derived from cereal grains, other seeds used as food, dairy products such as cheese and milk powders, dried fruits, dried and smoked meats and nuts. As well as their possible health significance, the presence of insects and insect excrete in packaged foods may render products unsaleable, causing considerable economic loss, as well as reduction in nutritional quality, production of off-flavours and acceleration of decay processes due to creation of higher temperatures and moisture levels.

Early stages of infestation are often difficult to detect; however, infestation can generally be spotted not only by the presence of the insects themselves but also by the products of their activities such as webbing, clumped-together food particles and holes in packaging materials. Unless plastic films are laminated with foil or paper, insects are able to penetrate most of them quite easily, the rate of penetration usually being directly related to film thickness.

In general, thicker films are more resistant than thinner films, and oriented films tend to be more effective than cast films. The looseness of the film has also been reported to be an important factor, loose films being more easily penetrated than tightly fitted films. Generally, the penetration varies depending on the basic resin from which the film is made, on the combination of materials, on the package structure, and of the species and stage of insects involved. The relative resistance to insect penetration of some flexible packaging materials is as follows:

— *excellent resistance:* polycarbonate; poly-ethylene-terephthalate;

— *good resistance:* cellulose acetate; polyamide; polyethylene (0.254 mm); polypropylene (biaxially oriented); poly-vinyl-chloride (unplasticised);

— *fair resistance:* acrylonitrile; poly-tetra-fluoro-ethylene; polyethylene (0.123 mm);

— *poor resistance:* regenerated cellulose; corrugated paper board; kraft paper; polyethylene (0.0254 - 0.100 mm); paper/foil/polyethylene laminate pouch; polyvinylchloride (plasticised).

Some simple methods for obtaining insect resistance of packaging materials are as following:

— select a film and a film thickness that are inherently resistant to insect penetration;

— use shrink film over-wraps to provide an additional barrier;

— seal carton flaps completely.

Rats and mice carry disease-producing organisms on their feet and/or in their intestinal tracts and are known to harbour salmonella of serotypes frequently associated with food-borne infections in humans. In addition to the public health consequences of rodent populations in close proximity to humans, these animals also compete intensively with humans for food.

Rats and mice gnaw to reach sources of food and drink and to keep their teeth short. Their incisor teeth are so strong that rats have been known to gnaw through lead pipes and unhardened concrete, as well as sacks, wood and flexible packaging materials. Proper sanitation in food processing and storage areas is the most effective weapon in the fight against rodents, since all packaging materials apart from metal and glass containers can be attacked by rats and mice.

REFERENCES

Cox, Peter A., *The Larger Rhododendron Species*. Portland: Timber Press, 1990.

Hay, Roy & Patrick M. Synge *The Colour Dictionary Of Garden Plants. With House and Greenhouse Plants.*, London: Bloomsbury Books, 1991.

Synge, Patrick M., *A Diversity In Plants*. London: Geoffrey Bles, 1953.

Wyman, Donald, *Wyman's Gardening Encyclopedia*. NY: Macmillan, 1972

9

Handling and Transportation of Horticultural Products

In agriculture, post-harvest handling is the stage of crop production immediately following harvest, including cooling, cleaning, sorting and packing. The instant a crop is removed from the ground, or separated from its parent plant, it begins to deteriorate. Post-harvest treatment largely determines final quality, whether a crop is sold for fresh consumption, or used as an ingredient in a processed food product.

The most important goals of post-harvest handling are keeping the product cool, to avoid moisture loss and slow down undesirable chemical changes, and avoiding physical damage such as bruising, to delay spoilage. Sanitation is also an important factor, to reduce the possibility of pathogens that could be carried by fresh produce, for example, as residue from contaminated washing water. After the field, post-harvest processing is usually continued in a packing house. This can be a simple shed, providing shade and running water, or a large-scale, sophisticated, mechanised facility, with conveyor belts, automated sorting and packing stations, walk-in coolers and the like.

In mechanised harvesting, processing may also begin as part of the actual harvest process, with initial cleaning and sorting performed by the harvesting machinery. Initial post-harvest storage conditions are critical to maintaining quality. Each crop has an optimum range for storage temperature and humidity. Also, certain crops cannot be effectively stored together, as unwanted chemical interactions can result. Various methods of high-speed cooling, and sophisticated refrigerated and atmosphere-controlled environments, are employed to prolong freshness, particularly in large-scale operations.

Regardless of the scale of harvest, from home garden to industrialised farm, the basic principles of post-harvest handling for most crops are the same:

— handle with care to avoid damage (cutting, crushing, bruising)
— cool immediately and maintain in cool conditions
— cull (remove damaged items)
— cool immediately and maintain in cool conditions
— cull (remove damaged items)

POST-HARVEST HANDLING

Packhouses: After product is harvested, it should be protected to prevent any cross-contamination. Packing facilities should be cleaned and well maintained to reduce the introduction of harmful microorganisms to product. Some growers move product from the field in large bins, which are taken to the packhouse for selection, grading, and repacking. No matter what method of packing is used, care must be taken with product. Cartons and other empty containers should not be stored uncovered in fields, because dust and animals could contaminate product.

Rigs and utensils used for packing should be cleaned and monitored daily. Packhouses, whether open or enclosed, should be cleaned and protected to deter pest entry and possible product contamination. Harvest storage facilities, containers, or bins should be cleaned regularly, based on a set programme. Poor sanitation in the packhouse could lead to the formation of biofilms. Biofilms are layers of bacteria that attach to surfaces like stainless steel and plastic, and also attach to each other with the help of polymeric materials.

The biofilms trap other bacteria, debris and nutrients. Poor sanitation programmes cause biofilms to build up and become established. Nonpathogenic and pathogenic bacteria can form biofilms. Organisms in the film tend to be resistant to cleaners and sanitisers, as well as heat treatment. Thus, in the food industry a sound sanitation programme is needed so that biofilms do not become entrenched on food contact surfaces. Good sanitation practices enhance a company's food-safety programme. An important step is to provide training in sanitation to a wide base of employees, even those outside the sanitation department. In daily sanitation programmes, seven steps could be followed to ensure clean equipment:

— Dry cleaning to remove gross debris from equipment and floors;
— Prerinsing to remove debris from surfaces;
— Using soap and scrubbing equipment on surfaces and floors;
— Postrinsing to remove soap;
— Removing standing water and reassembling equipment if necessary;
— Inspecting cleaned area and recleaning with detergent if necessary;
— Sanitising equipment and floors in high-care facilities.

Many companies use chlorine-based cleaning chemicals. For effective cleaning, some chemicals could require a specific residence time on equipment. Sanitisers include halogen-based compounds and quaternary ammonium-based compounds. For a review on food plant cleaning and sanitising, the reader is referred to Redemann. Pestmanagement could prevent product contamination, recall, and other loss of productivity. Acompany should have a preventive programme of pest control so that problems never develop.

The packhouse manager should assess the risk in the plant to determine the level of prevention needed. Completely closed packhouses may have sophisticated pest control programmes using a certified contract company. Windows and air vents should be screened and facilities kept free of debris. Any pest-control programme implemented should be monitored and documented regularly to protect product.

Water Flumes: Where water is used to transport produce from one part of the packhouse to the next, the quality of that water will determine the quality and safety of the product being packed. Primary water for rinsing to remove dirt may be of agricultural grade. However, potable water should be used for subsequent rinsing steps. Fresh-cut rinsing operations may use automated chlorination and acidification systems to control and monitor water quality. Flume channels used for sorting and grading product should be cleaned according to a planned sanitation programme, thus preventing the build-up of debris from recycled water and avoiding product contamination. Where water is recycled, it should be treated to reduce the build up of microorganisms.

Many companies use chlorine or other disinfecting chemicals to control the microbial load. Packers and processors use chlorine dioxide, chlorine (hypochlorous acid), UV light, ozone, hydrogen peroxide, peroxyacetic acid, and other sanitisers. The standard sanitiser used in packhouses in the United States is chlorine. This could be in the form of sodium hypochlorite granules, tablets, or liquid. Water has been demonstrated to enter tomatoes through the stem end in a water dunk tank.

This may be caused by the temperature differential of the produce item harvested in the field and the cooler water in the packhouse. At intake, any microorganisms in the water would also enter the produce item. Thus, it is imperative that any wash water in the plant be potable water. Any pathogens in water could become the source of a foodborne outbreak. A multi-State outbreak of Salmonella enterica Serotype Newport infection was linked to mango consumption in December 1999. Traceback of the implicated mangoes led to a single Brazilian farm, where hot water treatment was identified as a possible point of contamination.

Hot water treatment was a new process introduced to prevent importation of an agricultural pest, the Mediterranean fruit fly. Contaminated water caused product contamination. This outbreak highlighted the potential global health impact of foodborne diseases and newly implemented food processes and the vulnerability of product placed in contaminated water.

Employees: Packhouse employees should be trained in safe food-handling practices. Product safety of final consumer packs is directly influenced by the handling practices in the packhouse. Employees should receive training on the proper use of toilets, hand-washing procedures, use of protective clothing, and headgear to avoid product contamination. Many foodborne outbreaks have been linked to a sick employee who transferred pathogens via the fecal-oral route. Employees shedding pathogens in diarrhea may not wash their hands adequately before handling food. This is one possible route of contamination of fresh produce.

STORAGE AND DISTRIBUTION

Refrigeration temperatures in storage and distribution are crucial to maintaining product quality. These temperatures also reduce the proliferation of human pathogens if they are present on produce items. Refrigeration units are thought to spread mold throughout warehouses, and routine servicing of air filters and refrigeration systems is required. As cold air systems blow mold spores into the air, there is also the risk that pathogens may be spread along with the spores from one pallet to the other.

Pest control programmes are necessary at any storage facility. A basic rodent control programme would reduce the presence of pests that harbour harmful microorganisms forming a potential hazard to food. A contract company or trained individual should monitor traps or bait stations regularly and document any pest activity observed. Vehicles and containers used to transport fresh produce could also be sources of potential contamination. Vehicles used to transport fresh produce should be clean and free of odours, dirt, and debris before loading.

The ideal situation would be to use dedicated containers and vehicles for each application; that is, produce, meat, or refuse should be transported by separate means. The produce containers and vehicles should also be cleaned routinely as part of Good Hygienic Practices (GHPs) to prevent contamination between loads. Good hygienic and cleaning practices ensure product safety when loading or during inspections. The temperature of transport would also determine the potential for growth of pathogens. Thus, refrigeration temperatures are used to transport many produce items. The cold temperature helps to preserve product quality as well as safety.

FRESH-CUT FRUIT AND VEGETABLE INDUSTRY

The fresh-cut fruit and vegetable industry in the United States grew from supplying quick-serve restaurant chains to providing consumer-size convenience products. From the 1980s to the 2000s, bagged salads and other convenience items experienced tremendous growth in demand. Sales of bagged salads topped $3 billion in 2004, while whole peeled carrots reached about $1 billion. Fresh-cut processors have worked with growers, retailers, and food service operations to ensure raw material and finished product are handled safely.

When fresh produce is washed, cut, and sliced, the natural defence mechanisms on the plant material are removed. The high level of handling increases the potential for product to be contaminated by microorganisms in the environment. Because there is no kill step in fresh-cut processing, using HACCP is imperative in maintaining food safety standards with preventive programmes. Growers implement Good Agricultural Practices (GAPs) reducing risks during growing and harvesting. Retailers and food service operations implement the principles of HACCP to manage food safety in their operations.

In the United States, the International Fresh Cut Produce Association (IFPA) published Food Safety Guidelines for fresh-cut food processors. Guidance documents produced include a model HACCP plan, best practice guidelines for activities, a model food allergen plan, and a Sanitary Equipment Buying Guide and Development Checklist. The IFPA has also developed a GAP programme in conjunction with the Western Growers' Association. The industry is thus equipped with proactive guidance to produce safe food. The primary cause of foodborne illness in the United States is thought to be the mishandling of food during preparation in food service operations or in the home.

Statistics collected by the Center for Disease Control and Prevention (CDC) provide a strong link between foodborne illness outbreaks and the end users in restaurants and retail. Consumers could be a source of fresh produce contamination in retail outlets. Consumers touch fruit and vegetables as they make a decision on whether to purchase product. If a person's hands are contaminated because of improper hygiene, this product could be affected.

The consumer may also place bare fruits and vegetables into shopping carts, which are generally unwashed. Cross-contamination by microorganisms present in the shopping cart onto fresh produce items could present a food safety hazard, especially if the cart was used to transport meat, fish, or poultry. The final step of bagging produce items could present a food safety hazard if fruits and vegetables are placed in the same bag as raw meat or fish. When the consumer places grocery bags in the car, vehicle temperature and time to cooling determine the potential for present pathogens to multiply.

Retail operations and restaurants should take a proactive role in carrying out a risk assessment of their operations to understand the different types of hazards. When these potential problems are identified using the principles of HACCP, they may be controlled. Salads and juices are sometimes prepared in the fresh produce section of a retail operation. Workers should be trained in good food-handling practices to ensure food quality and safety are maintained:

— Ready-to-eat foods should be kept at safe temperatures.
— Food contact surfaces in food-preparation areas should be cleaned and sanitised regularly.
— Display counters and shelves should be cleaned regularly to prevent crosscontamination.
— Water used to wash produce should be potable, because poor-quality water is a major source of contamination.
— Colour-coded chopping boards should be used to distinguish boards for cutting produce from those used for cutting meat.
— Employees bagging product at the cashier's desk should be trained to avoid cross-contamination by placing raw meat in separate bags from fresh produce.

— Regular hand washing should be enforced to prevent product contamination.

All employees in a retail operation should thus have a basic knowledge of food safety and their responsibility of protecting the public. Outreach and education of the consumer would be the most effective means of explaining the dangers of mishandling food to the ultimate customers. Retailers may also assume the role of educating the public so the food-safety message reaches into people's homes. Brochures, videos, and posters at display counters have been used effectively for consumer education in retail operations.

Modern technology and a bit of creativity on the part of retailers could go even further to enlighten the public and promote safe handling of food. Restaurant chains represent one important point in the handling of fresh fruits and vegetables before consumption. Food-service operations have been linked to several foodborne outbreaks involving fresh produce. Adequate training of food-service employees and implementation of a HACCP programme to control food-safety in kitchens and food-display counters are the best ways to control food handling.

The "Serve Safe" programme developed by the International Food Safety Council of the National Restaurant Association Educational Foundation has become the industry standard in food safety training. Serve Safe is accepted in almost all U.S. jurisdictions that require employees' certification. The Serve Safe programme provides accurate, up-to-date information for all levels of employees on all aspects of handling food, from receiving and storing to preparing and serving. The International Food Safety Council promotes food-safety education to the restaurant and food service industry, and also conveys the industry's food-safety commitment to the public.

The Serve Safe programme has introduced a number of control measures for food-service operations similar to those mentioned above for retailers handling minimally processed product. In the United States, food service operations may also be subject to food-safety inspections, depending on the laws in each State. Only a very strict food-safety programme in food-service operations would reduce the risk of pathogen contamination and illness to the ultimate consumer.

CONSUMER HANDLING

Fruit and vegetables, besides being perishable items, could be the source of mishandling by consumers, eventually leading to foodborne illness. Consumers sometimes mishandle produce by cross-contamination with meat items being placed in the same bag or cart. Temperature abuse in a hot car would promote the growth of microorganisms, if present. In the home, food-safety practices such as hand washing before handling fresh produce, or after handling meat, may not be observed.

Most consumers store produce in the refrigerator; however, some items are stored at room temperature. Room-temperature storage is good for tomatoes, bananas, and unripe climacteric fruits. However, refrigeration extends the freshness and slows bacterial growth if a produce item contains harmful microorganisms. The home refrigerator may also harbour harmful organisms if it is not cleaned routinely. Meat drippings on shelves could cross-contaminate fresh produce placed on the same shelves. Consumers do not generally have a cleaning schedule for refrigerators, but this is a critical point in home hygiene in the kitchen.

Refrigerator surfaces should be cleaned and sanitised to remove potential pathogens every two weeks, or more often, depending on contamination. Another reservoir for pathogenic microorganisms is the kitchen sink. Consumers may place fresh produce items in the sink without washing or sanitising the area. This causes cross-contamination from items previously placed in the sink. Sinks should be cleaned and sanitised before being used to wash fresh fruits and vegetables. In addition, mixing of utensils would invite cross-contamination, leading to foodborne illness.

Consumers may not always wash fruits and vegetables, but even the simplest washing with running water is sufficient to cause one log_{10} CFU/g reduction in microbes. Fresh-cut fruit and vegetables should always be stored in refrigeration, first to extend shelf life, and then to reduce the potential growth of pathogens, which are usually mesophilic organisms. Chopping boards carry microorganisms in the form of biofilms if they are not cleaned and sanitised thoroughly. Consumers should implement good cleaning programmes when using chopping boards for fresh produce, and preferably use different utensils than those used for meat.

The presence and survival of human pathogens in unpasteurised juices, sprouts, melons, lettuce, and berries is known. These commodities have been implicated in a number of foodborne illness outbreaks. Typical methods of washing and sanitising produce are not effective in removing pathogens, and increasing the knowledge of sources of microbial contamination requires more research. Thus, scientists have tried to identify sources of human pathogen contamination and develop interventions to prevent contamination and remove or inactivate pathogens on fresh and minimally processed produce. These studies provide a foundation for the development of treatments to assure the microbiological safety of fresh produce.

However, experimental washing with hydrogen peroxide resulted in a 3-log (99.9 percent) reduction. Annous et al. demonstrated the inability of brush washing to decontaminate apples in field tests. It was thought that bacteria in inaccessible sites, growth of bacteria in skin punctures, and possible infiltration of bacteria into calyx and core tissues prevented their destruction. In work with apples, mature and immature fruit from various locations were examined for evidence of internal or external bacterial contamination. A decaycausing fungus, Glomerella cingulata, permitted growth of E. coli O157:H7 in inoculated apples, probably because of a decrease in acidity that resulted from fungal growth. This clearly demonstrates the potential risk of using decayed apples for production of unpasteurised cider.

Sapers et al. researched the microbiology of fresh and fresh-cut cantaloupe, providing knowledge about attachment and survival of bacteria, including surrogates of human pathogens on external melon surfaces. They studied the efficacy of washing treatments in decontaminating cantaloupe melons, and effective treatments for extending the shelf-life of fresh-cut cantaloupe. Cantaloupe melons were artificially contaminated with nonpathogenic E. coli and Salmonella stanley, a human pathogen, and survival of these bacteria on melon rind during washing and their transfer to the melon flesh during fresh-cut processing was measured.

Researchers found that chlorine and hydrogen peroxide solutions became progressively less effective in reducing contaminant populations as the time between contamination and washing increased. Surviving bacteria were transferred to the melon flesh and growth occurred. Chlorine dioxide is one sanitiser approved by the U.S. Food and Drug Administration (FDA) and the Environmental Protection Agency (EPA). Chlorine dioxide is said to be as powerful as peroxyacetic acid and more economical. It has less of a negative impact on the environment than quaternary ammonium salts, chlorine, or bromine.

Ozone and peroxyacetic acid have been introduced to decrease the microbial content of fresh-cut produce. Chlorine bleach and bromine create carcinogenic trihalomethanes that enter the environment through drains. Chlorine dioxide is also thought to destroy biofilms and prevent their formation. Chlorine continues to be the most widely used sanitiser in the produce industry in the United States; however, certain countries in the European Union have banned its use. Heat pasteurisation is one innovation receiving a great deal of attention in disinfecting fresh-cut apples and melons.

The outside of the fruit is pasteurised before cutting, thus reducing the entry of surface microorganisms to the flesh. In fresh-cut processing operations, ATP bioluminescence is used to monitor food contact surface contamination. It is a rapid means of informing sanitation crews if bacteria and organic material remain on food contact surfaces. This method of testing is, however, very expensive and finds application mainly in large processing operations. US Food Safety Programmes: GAP, GMP, and HACCP. Use of GAP, GMP, and HACCP is voluntary in the fresh produce sector. However, retailers have demanded that their suppliers implement structured food-safety programmes under various schemes and these practices are now subject to auditing by a third party.

Good Agricultural Practices (GAPs) describe preventive measures implemented in farming operations to reduce product contamination. The U.S. government in 1998 introduced a "Guide to Minimise the Microbial Contamination of Fresh Fruit and Vegetables". The major risk factors identified as needing control included water, manure, worker health and hygiene, sanitary facilities, pest control, and traceback. GAPs provide guidance for food-safety practices in the field. Implementing GAPs provides some assurance to the retailer that product is safe from the farmer's gate.

A small number of foodborne illnesses have been associated with contamination at the farm through an ill harvester, or use of unsanitary water. Current GMPs were made law in U.S. food-processing and handling facilities and this code has been reviewed. The code provides guidance on the safe handling of product in buildings used for food production. In the fresh produce sector, packhouses should follow guidance set out in cGMPs in order to put out wholesome food product.

The principles of HACCP, a system developed by the NASA space programme in the 1950s, are now used as the basis for risk assessment and initiating food-safety measures throughout the industry. Although most fresh produce farming and packing operations do not have critical control points, the HACCP programme forces operations to go through each area in a packhouse, conducting a risk assessment. The assessment may also indicate where GMPs are failing and help in improving GMPs. The use of GAPs, GMPs, and HACCP in the fresh fruit and vegetable industry provide the basic framework for safe products for the consumer.

DISTRIBUTION AND RETAIL

To aid the integration of the supply chain from the source to the end consumer, automated systems are being developed. These data-based systems would provide complete and constant visibility of product, package, purchasing, and distribution of food. Automated warehouse management systems using robotics could increase accuracy of the picking and assembling of pallets for dispatch. With laws restricting the weight of product lifted at food plants, automated systems could also aid manufacturers in compliance with labor laws. As customer demands for fresh product increases, smaller, more frequent deliveries will be required.

Automated warehouse management systems would again improve accuracy of product rotation and picking for small deliveries. Automated systems would improve inventory control and real-time tracking of products that pass through a warehouse. The advent of barcoding has revolutionised the way product movement is managed. A unique barcode is prepared for product harvested on the farm. This barcode stores information on the grower, farm lot number, day harvested, and crop harvested.

A barcode placed on a bin of fruit after harvest could then be scanned at the packhouse at intake. This code stays with the bin until it is repacked into consumer size items. If product were stored before repackaging, the warehouse management system would assist in stock rotation based on information received from the barcode. When product is repacked into consumer size packages, a new barcode is placed on the product to assist in tracing finished product back to the field in case of an emergency.

The Uniform Code Council (UCC) takes a global leadership role in establishing and promoting multi-industry standards for product identification and related electronic communication. The purpose of CC/EAN-128 is to establish a standard way of labeling a package with more information than just a product code. It provides supplemental information such as batch number and "use before" dates. The symbology for this is the UCC/EAN-128, which can be scanned to capture automatically the details related to the location. International standards in barcoding include the use of a Global Location Number or GLN. This is a number that identifies any legal, functional, or physical location within a business or organisational entity such as:

— *Legal entities*—whole companies, subsidiaries, or divisions such as supplier, customer, bank, forwarder;
— *Functional entities*—a specific department within a legal entity, such as the accounting department;
— *Physical entities*—a particular room in a building, such as a warehouse or warehouse gate, delivery point, or transmission point.

Such GLNs are governed by strict rules to guarantee that each one is unique worldwide. They always have the following features:

— They are numeric.
— They have a fixed length of 13 digits, which must be processed in their entirety.
— They end with a check digit.

The GLNs can be presented in barcode format and physically marked onto trade units to identify the parties involved in the transaction (buyer, supplier), for example (1) transport units (consignor, consignee), (2) physical locations (place of delivery, place of departure). X-ray technology is another method with great potential in identifying physical contamination in fresh fruits and vegetables.

The widely used metal detectors are not able to detect wood chips, plastic, and other nonmetal contaminants. Thus, machine vision systems could help in eliminating foreign contaminants in food. The cost of purchasing equipment and adapting it to fresh-produce operations far exceeds the benefits from using this technology. Although some in the produce industry are still grappling with the introduction of barcoding systems for traceability programmes, other sectors in the food industry have moved on to the new technology of radio-frequency identification (RFID) systems, a technology that uses grain-sized computer chips to track items at a distance.

Each tiny chip is connected to an antenna that picks up electromagnetic energy beamed at it from a reader device. When it picks up the energy, the chip sends back its unique identification number to the reader device, allowing the item to be remotely identified. "Spy chips," as they are sometimes called, can beam back information anywhere from a few inches to 30 ft away. This RFID will be used in the future to trace product. Retailers such as Walmart have started demanding this technology for packaged goods. Within time, this technology could be used to assist in stock rotation and distribution of fresh produce. Radio frequency identification is not necessarily better than barcodes.

The two are different technologies and have different applications, which sometimes overlap. The big difference between the two is that barcodes are a line-of-sight technology. In other words, a scanner has to see the barcode to read it. This means that people usually have to orient the barcode towards a scanner for it to be read. Radio frequency identification, by contrast, does not require line of sight. RFID tags can be read as long as they are within range of a reader. All existing RFID systems use proprietary technology, but many of the benefits of tracking items come from tracking them as they move from one company to another and even one country to another.

POST-HARVEST LOSSES

Poor planning or inaccurate production and market information may lead to over-production of certain fruits or vegetables which cannot be sold in time. This situation occurs most frequently in areas where transportation and storage facilities are inadequate. Produce may lie rotting in production areas, if farmers are unable to transport it to people who need it in distant locations. Although many people believe that postharvest losses are higher in less industrialised countries, this generalisation may not be always true.

Higher losses may occur in developing countries for lack of good facilities and technologies. However, post-harvest losses may be lower in less urbanised developing countries, where the products need to be transported a shorter distance to market, and there is a shorter time lag period between harvesting and consumption. Consumers in such countries are also more likely to buy lower-grade products. In industrialised countries, on the other hand, produce is shipped for longer distances, the marketing system is more complicated, and consumers demand good quality when they buy food.

All these factors contribute to greater postharvest losses which may not be fully compensated by better facilities and technologies. For instance, Bourne cited a survey report that losses of produce after arrival in New York City were 1.7 14.2%. This includes only losses in wholesale markets, retail outlets and by consumers. It is safe to say that post-harvest losses occur in every country, but the magnitude and major causes of losses and the effective remedial methods differ greatly from one country to another.

In order to solve specific problems in specific areas effectively and economically, a comprehensive knowledge of the nature of post-harvest losses, a grasp of various kinds of technologies, and a tactical selection of strategies is necessary. Post-harvest losses have many different forms. The most obvious loss is a quantitative loss, such as reduced weight and partial or total waste of a product due to decay or senescence. A less discernible loss is a qualitative one, such as deterioration in texture, flavour or nutritional value. Other forms include loss of viability of propagules and reduction in monetary value due to reduced prices.

Post-harvest losses can occur in the field, in packing areas, in storage, during transportation, and in wholesale or retail markets. Severe losses occur because of poor facilities, lack of know-how, poor management, market dysfunction, or simply the carelessness of farmers or workers. Postharvest losses may also occur at consumers' homes, in the kitchen or on dining tables. However, losses after produce has left the retail market are generally difficult to control by agricultural means.

The causes of post-harvest losses are:

— *Mechanical Injury*: Owing to their tender texture and high moisture content, fresh fruits and vegetables are very susceptible to mechanical injury. Poor handling, unsuitable containers, improper packaging and transportation can easily cause bruising, cutting, breaking, impact wounding, and other forms of injury.

— *Physiological deterioration*: Fruit and vegetable cells are still alive after harvest, and continue their physiological activity. Physiological disorders may occur due to mineral deficiency, low or high temperature injury, or undesirable atmospheric conditions, such as high humidity. Physiological deterioration can also occur spontaneously by enzymatic action leading to overripeness and senescence, a simple aging phenomenon.

— *Parasitic diseases*: High post-harvest losses are caused by the invasion of fungi, bacteria, insects and other organisms. Microorganisms attack fresh produce easily and spread quickly, because the produce does not have much of a natural defense mechanism, and has plenty of nutrients and moisture to support microbial growth. Post-harvest decay control is becoming a more difficult task, because the number of pesticides available is falling rapidly as consumer concern for food safety increases.

PRINCIPLES OF TRANSPORTATION

Transportation is often the most costly factor in the marketing channel, and for airfreighted export crops the cost of transportation may exceed the cost of production. The method of transportation for fresh fruit and vegetables is determined by distance, perishability and the value of the product. Whatever the method used, the principles of transport are the same:

— Loading and unloading should be as careful as possible.

— Transmit times should be as short as possible.

— The product should be well protected in relation to its susceptibility to physical injury.

- Jolting and movement should be reduced as much as possible.
- Overheating should be avoided.
- Water loss by the produce should be restricted.
- The required conditions of preservation should be obtained and maintained constantly, regarding in particular temperature, relative humidity and air circulation.

REFERENCES

Cassell, Clark C., *The World is Your Market - An Export Guide for Small Business,* Edited by William Delphos, Braddock Communications, Washington, DC, 1990.

Hardenburg, R. E., A. E. Watada, C. Y. Wang, *The Commercial Storage of Fruits, Vegetables, and Florist and Nursery Stocks,* HBK-66. USDA, ARS, September 1986.

Nicholas, C. J., *Export Handbook for U.S. Agricultural Products,* HBK-593. USDA, AMS, March 1985.

U.S. Department of Agriculture, Foreign Agricultural Service, *Food & Agricultural Export Directory 1991.* MPN-1497, September 1991.

10

Storage of Fruits and Vegetables

Generally, storage facilities are linked or integrated to packinghouses or other areas where there is a concentration of product. However, often storage can also be undertaken on-farm, either naturally or in specifically designed facilities. Even under conditions of mechanical refrigeration, location and design have an impact on system operations and efficiency. First, climate is an important factor for the location of the storage facility. For example, altitude reduces temperature by 10 °C for every 1000 meters of elevation. It also increases overall efficiency of the refrigeration equipment by facilitating heat exchange with ambient temperature, thereby reducing energy costs. Shading particularly of loading and unloading areas reduces thermal differences between field and storage temperatures.

Building design is an important factor to be taken into consideration. For example, a square shaped floor perimeter is thermally more efficient than a rectangular one. The roof is the most important part of the structure. This is because it has to protect produce from rain and radiant heat. Its slope should allow easy fall off of rainwater; its dimensions should exceed the perimeter of the building to protect walls from the sun and provide a dry area around the building in rainy weather. Floors should be of concrete, isolated from soil humidity, and elevated to avoid penetration of water.

Doors need to be wide enough for mechanised handling. Storage facilities should be thoroughly cleaned before filling. This includes brushing and washing of walls and floors to eliminate dirt and organic debris that could harbour insects and diseases. Before product is placed in the storage room, inspection and presorting should be undertaken. This is in order to remove all potential sources of contamination for the remaining load. Product should be stacked in such a way that there is free circulation of air. During storage, it should also be possible to carry out quality control inspections. If the storage facility becomes full during a long harvest period, it needs to be organised around the principles of the system "first in first out".

DIFFERENT TYPES OF STORING

As a rule, there are many ways of storing a product. The length of storage time can be longer in specifically designed structures. With refrigeration and controlled atmospheres, storage periods can be even longer. The technology utilised depends on whether the benefits outweigh the costs.

Natural Storage

This is the most rudimentary system and is still in use for many crops. For example, roots and tubers. Crops should be left in the soil until preparation for the market. This is similar to how citrus and some other fruits are left on the tree. Although storing products under natural conditions is widely practiced, it leaves them exposed to pests and diseases as well as to adverse weather conditions. This can have a detrimental effect on quality.

Another method widely used is field storage in heaps. This method ensures that produce is free from soil humidity and is protected from the weather with a tarpaulin, straw, or plastic materials. It is a low cost alternative for bulky crops that require large buildings. For example, potato, onions, pumpkins, sweet potato etc. Field storage in bins is a more recent variation where a pair of them is left in the field. It has the additional advantage of making it possible to undertake mechanical handling later.

Natural Ventilation

Amongst the wide range of storage systems, this is the most simple. It takes advantage of the natural airflow around the product to remove heat and humidity generated by respiration. Buildings providing some form of protection from the external environment and with gaps for ventilation can be used. Produce can be placed in bulk, bags, boxes, bins, pallets etc. Although simple, some key concepts need to be taken into account for the efficient operation of this system.

1. Differences in internal temperature and relative humidity conditions compared to conditions externally, need to be minimal. What this means is that this system can only be used with crops that store well under natural conditions such as potato, onions, sweet potato, garlic, pumpkins, etc.
2. For adequate ventilation, openings need to be wide. This means they need to be fitted with screens to keep animals, rodents, and pests out.
3. As with any other type of fluid, air follows the path of least resistance. This means that if product is stored in a compact mass, air will circulate to remove heat and gases which have accumulated as a result of respiration. Efficient ventilation requires adequate space. However, this reduces storage capacity.
4. Hot and humid air rises within the storage facility. If no ventilation gaps exist, this leads to the build up of hot and humid areas which in turn affects the quality of stored goods. This presents the ideal conditions for the development of disease.

Within certain limits, it is possible to take advantage of natural changes in temperature and relative humidity. This can be achieved by selectively opening and closing the storage ventilation. At noon, ambient temperature and relative humidity are higher and lower, respectively. However, at night the opposite happens. To reduce temperature of stored products, buildings should be left open when external air temperatures are lower. Internal relative humidity can also be managed in a similar way.

External conditions constantly change, even during the same day. However, in comparison to air, stored mass is slower to gain and release heat. In order to handle this efficiently, internal and external electronic sensors for temperature and relative humidity are required. In addition to this, although crops suitable for this type of storage have low respiratory rates, some ventilation may be required. This is in addition to the automated opening and closing schedules.

Forced Air Ventilation

Heat and gas exchange can be improved provided air is forced to pass through the stored product. This system allows for more efficient utilisation of space for bulk storage. Air conducts run under a perforated floor and air is forced through the product. Again, as air follows the least resistance path, loading patterns as well as fan capacity and conduct dimensions should be carefully calculated. This is to ensure that there is uniform distribution of air throughout the product.

Removable perforated ducts can be used for storage space when there are no products in storage. Fan selection is the most critical factor and specialised personnel should design the system based on volume and number of air changes per unit of time required. The latter is a function of respiratory rates of products to be stored. Static pressure or resistance to the airflow by conducts and stored mass should be considered. Ideally, sensors reacting to the internal/external ambient relationship should control the system. If closed, internal air circulation only occurs. On the other hand, if opened internal atmosphere is replaced by ventilation. A partial opening produces a mix of internal and external air to reach the desired combination of temperature and relative humidity.

Controlling Temperature

Controlling temperature is one of the main tools for extending postharvest life: low temperatures slow product metabolism and the activity of microorganisms responsible for quality deterioration. As a result, reserves are maintained with a lower respiration rate, ripening is retarded and vapour pressure between product and ambient is minimised, reducing water loss. These factors contribute towards maintaining freshness by reducing the rate at which quality deteriorates and the nutritional value of the product is preserved. A refrigerated room is a relatively airtight and thermally insulated building.

The refrigeration equipment should have an external escape outlet to release externally the heat generated by the product. Refrigeration capacity of the equipment should be adequate to extract the heat generated by crops with a high respiration rate. It is also important to precisely control temperature and relative humidity conditions inside the refrigerated storage environment. Refrigerated space depends on the maximum storage volume.

Other factors to be considered include walkways and aisles to handle the product mechanically and the additional space to ensure uniform distribution of cold air. It is not uncommon to find that produce occupies only 75-80% of total surface area. Chamber height depends on product and stacking pattern: three meters for hand stacking but more than six may be required if forklifts are utilised. Refrigerated rooms can be made with concrete, metal, wood, or other materials. All external surfaces should be thermally insulated, including the floor and ceilings.

Type and thickness of insulation material depends on building characteristics, produce to be stored and the difference in temperature required between external and internal conditions. Polyurethane, expanded polystyrene, cork and other such materials can be used as insulation materials. A vapour barrier should be placed on the warm side of the insulation material. Mechanical refrigeration has two main components: the evaporator, inside the storage area and the condenser which is outside connected by tubing filled with refrigerant.

Normally, both elements are finned coils made of high thermal conductivity materials and integrated to a fan. This facilitates heat exchange. An evaporator is placed in the upper part of one of the walls forcing cold air to flow parallel to the ceiling. Returning air is forced past the evaporator transferring to the coil the heat extracted from the product. A refrigerant absorbs this heat as it changes to gas, cooling the air, which is forced again into the room as cold air. The refrigerant is transported as gas to the condenser where under the pressure provided by a compressor, it is transformed again into the liquid form.

The internal heat is then released outside. With this repeated cycle, the system behaves like a pump—heat is extracted from the stored product and then released outside. Another key aspect of the mechanical refrigeration system is the expansion valve, which regulates the evaporation and flow of refrigerant. Ammonia and Freon gas are the most widely used refrigerants. However, they are now being replaced by more environmentally friendly products. In addition to design and consideration of building materials, to gain maximum benefit from refrigeration the following conditions need to be met: refrigeration capacity needs to be adequate—this is in order to extract respiration heat from the product as well as conductive heat; convective heat gains, and the heat produced by equipment. Every crop has an optimal combination of temperature and relative humidity for storage. In many cases, there are differences even within varieties.

Long-term Storage

Facilities for long-term storage of potato, onion, sweet potato etc, often include using a combination of forced air systems as well as heating and/or refrigeration equipment. Because these are crops that initially require a curing period, hot and humid air is introduced at the beginning. Later, the temperature is reduced with either through forced air cooling or natural ventilation. Adequate temperatures are obtained by mixing external and internal atmospheres and if required, the air is heated or refrigerated. In this way, the same building is used for both curing and storage—an important consideration in mechanised harvesting systems.

INTERMEDIATE MOISTURE FOODS (IMF)

Traditional intermediate moisture foods (IMF) can be regarded as one of the oldest foods preserved by man. The mixing of ingredients to achieve a given a_w, that allowed safe storage while maintaining enough water for palatability, was only done, however, on an empirical basis. The work done by food scientists approximately three decades ago, in the search for convenient stable products through removal of water, resulted in the so-called modern intermediate moisture foods. These foods rely heavily on the addition of humectants and preservatives to prevent or reduce the growth of microorganisms. Since then, this category of products has been subjected to continuous revision and discussion.

Definitions of IMF in terms of a_w values and moisture content vary within wide limits, and the addition of preservatives provides the margin of safety against spoilage organisms tolerant to low a_w. Of the food poisoning bacteria, Staphylococcus aureus is one of the organisms of high concern since it has been reported to tolerate a_w as low as 0.83-0.86 under aerobic conditions. Many of the considerations on the significance of microorganisms in IMF are made in terms of a_w limits for growth. However, microbial control in IMF does not only depend on a_w but on pH, Eh, F and T values preservatives, competitive microflora, etc., which also exert an important effect on colonising flora.

The application of IMF technology has been very successful in preserving fruits and vegetables without refrigeration in most Latin American countries. For instance, the addition of high amounts of sugar to fruits during processing will create a protective layer against microbial contamination after the heat process. The sugar acts as a water activity depressor limiting the capability of bacteria to grow in food.

Food products formulated under this concept are stable at room temperature without thermal processing and can be generally eaten without rehydration. Some processed fruits and vegetables are considered IMF foods. Advantages of IMF preservation are: Intermediate moisture foods have an a_w range of 0.65-0.90, and thus water activity is their primary hurdle to achieving microbial stability and safety.

IMF foods are easy to prepare and store without refrigeration. They are energy efficient and relatively cheap. They are not readily subject to spoilage, even if packages have been damaged prior to opening, as with thermostabilised foods, because of low a_w. This is a plus for many developing countries, especially those in tropical climates with inadequate infrastructure for processing and storage, and offers marketing advantages for consumers all over the world.

Disadvantages of IMF preservation are: Some IMF foods contain high levels of additives that may cause health concerns and possible legal problems. High sugar content is also a concern because of the high calorific intake. Therefore, efforts are been made to improve the quality of such foods by decreasing sugar and salt addition, as well as by increasing the moisture content and a_w, but without sacrificing the microbial stability and safety of products if stored without refrigeration. This may be achieved by an intelligent application of hurdles. Fruit products from intermediate moisture foods (IMF) appear to have potential markets. However, application of this technology to produce stable products at ambient temperature is limited by the high concentration of solutes required to reduce water activities to safe levels. This usually affects the sensory properties of the food.

COMBINED METHODS FOR FOOD PRESERVATION

Food preserved by combined methods (hurdles) remains stable and safe even without refrigeration, and is high in sensory and nutritive value due to the gentle process applied. Hurdle technology is the term often applied when foods are preserved by a combination of processes. The hurdle includes temperature, water activity, redox potential, modified atmosphere, preservatives, etc. The concept is that for a given food the bacteria should not be able to "jump over" all of the hurdles present, and so should be inhibited.

If several hurdles are used simultaneously, a gentle preservation could be applied, which nevertheless secures stable and safe foods of high sensory and nutritional properties. This is because different hurdles in a food often have a synergistic or additive effect. For instance, modified foods may be designed to require no refrigeration and thus save energy. On the other hand, preservatives could be partially replaced by certain hurdles in a food. Moreover, a hurdle could be used without affecting the integrity of food pieces or in the application of high pressure for the preservation of other foods. Hurdle technology is applicable both in large and small food industries.

In general, hurdle technology is now widely used for food design in making new products according to the needs of processors and consumers. For instance, if energy preservation is the goal, then energy consumption hurdles such as refrigeration can be replaced by hurdles that do not require energy and still ensure a stable and safe product. The hurdle effect is an illustration of the fact that in most foods several factors contribute to stability and safety. This hurdle effect is of fundamental importance for the preservation of food, since the hurdles in a stable product control microbial spoilage and food poisoning as well as undesirable fermentation.

Increasing consumer demand for fresh quality products is turning processors to the so-called minimally processed products (MP), an attempt to combine freshness with convenience to the point that even the traditional whole, fresh fruit or vegetable is being packaged and marketed in ways formerly reserved for processed products. According to these authors, the widely accepted concept of MP refrigerated fruits involves the idea of living respiring tissues. Because MP refrigerated products can be raw, the cells of the vegetative tissue may be alive and respiring, and biochemical reactions can take place that lead to rapid senescence and/or quality changes.

In these products, the primary spoilage mechanisms are microbial growth and physiological and biochemical changes, and in most cases, minimally processed foods are more perishable than the unprocessed raw materials from which they are made. The technology for shelf-stable high moisture fruit products (HMFP) is based on a combination of inhibiting factors to combat the deleterious effects of microorganisms in fruits, including additional factors to reduce major quality losses from reactions. In order to select a combination of factors and levels, the type of microorganism and quality loss from reactions that might occur must be anticipated.

Minimal processing may encompass pre-cut refrigerated fruits, peeled refrigerated whole fruits, sous vide dishes, which may include pre-heated vegetables and fruits, cloudy and clarified refrigerated juices, freshly squeezed juices, etc. All of these products have special packaging requirements coupled with refrigeration. These products, apart from special handling, preparation, and size reduction operations, might also require special distribution and utilisation operations such as Controlled atmosphere/Modified atmosphere/air flow rate/vacuum storage, computer controlled warehousing, retailing and food service, communications network, etc. HMFP fruits are less sophisticated than MPR fruits and should be priced lower when introduced commercially. Careful selection of these processes should of course be made to find the appropriate methods suited to a particular rural or village situation.

DRYING METHODS

Foods can be dried in an electric dehydrator, in the sun, in a solar dryer, or in a regular oven.

Dehydrator Drying

Dehydrator drying produces the best quality dried products, so it's not surprising that it's also the most popular drying method. Dehydrator drying also gives you greater flexibility than other methods because it does not depend on dry, sunny days or take over your oven. A variety of electric dehydrators are available for purchase. A dehydrator should have a heat source, a thermostat, and some method of air circulation. If you buy a dehydrator, follow the directions that come with it. Before you buy a new or used food dehydrator, check to see that it has all these features:

— Instruction manual.
— Thermostatically controlled temperature dial with settings between 130° and 150°F. (If you plan to dry meat jerky in your dehydrator, the dehydrator must be capable of maintaining a temperature of 145°. Contact the extension educator in your county for instructions on how to safely prepare meat jerky.)
— Fan or blower to distribute warm air evenly. Shelves made of stainless steel or food-grade plastic. (Galvanized screening is not food-safe.)
— Easy loading and unloading features.
— Outside cabinets made of hard plastic, aluminum, or steel. The highest quality dehydrator has double-wall construction with insulating material sandwiched between the walls to reduce the amount of heat lost during use.
— Enclosed heating element. Appropriate number of trays for your use. Most food dryers come with 4 to 10 food trays.
— Source of replacement parts.

Sun Drying

Sun drying works best when the temperature is in the 90s, the humidity is low, and air pollution levels are low. A major advantage to sun drying is its low cost. Drying trays, netting to protect against bugs, and food to dry are your only investments. Sun drying makes you dependent on the weather, however. If it is sunny one day and cloudy the next, you will have to finish drying by another method. That's because spoilage can occur while the drying food still has enough moisture for microbial growth. Also, you should bring the food inside on cool nights. Another disadvantage is time. What dries in 6 to 8 hours in an electric dehydrator may take 2 to 4 days in the sun.

Solar Drying

Solar drying is like sun drying, only better. The sun's rays collect in a solar box so that, compared with sun drying, drying temperature is higher and drying time is shorter. The shorter drying time gives microorganisms less chance to cause spoilage. If you do not want to buy or build a solar box, you can use the back window ledge of an automobile where the sun shines through. Crack the windows slightly to allow air flow so temperatures do not get too hot.

Cover the trays with netting to keep bugs out. Oven Drying You can use your oven to dry small amounts of food at one time. You'll have little or no investment in equipment and you won't have to depend on the weather. Although oven drying produces a safe, generally tasty product, don't expect top quality. Oven-dried food is more brittle and usually darker and less flavourful than food dried in a dehydrator. Another disadvantage of oven drying is its energy cost. Oven drying takes two or three times longer than drying in a dehydrator. Before drying in an oven, test the oven temperature with an oven thermometer for about 1 hour. Prop open the oven door as you would when actually drying fruit.

The oven should maintain a temperature of 130° to 150°F. If the oven cannot maintain a temperature in this range, you will not have high-quality dried food. If the oven is too hot, your food will begin to cook instead of dry. If it is too cool, your food may not dry fast enough and spoil instead. If you're new to drying, start with the fruits you like best. Think also about how you will use your dried fruits. Peaches or pears in a tangy stewed fruit? Apples or apricots for lunch box snacks? Berries to toss into muffin or cake batters? Most fruits are easy to dry High-quality fruits make the best dried products. Choose firm, fully ripe fruit that is heavy for its size.

Handle fruits gently and process them immediately because fruit ready for drying is very fragile. Use overripe or bruised fruits in other ways. Vegetables for drying should be fresh, tender, and just mature. Avoid immature vegetables because their colour and flavour tend to be weak or poor. Also avoid excessively mature vegetables, which are inclined to be tough, woody, or fibrous. For the best quality and nutrition, dry vegetables as soon as possible after harvest.

PREPARATION OF DRIED FOODS

Gently wash all fruits in cold water just before drying to remove dirt, bacteria, and insects. Thoroughly wash fruits that have skins you will not peel off, such as cherries and prunes. Do not soak fruit because extended soaking can cause nutrient loss and waterlog the fruit, which increases drying times. Remove fruit stems and peels. Peels may be left on some fruits, such as apples and peaches, but they may become bitter or discolour during drying. Core or pit the fruit and cut it into uniform halves, quarters, or slices. Trim away diseased or soft spots.

Wash vegetables in cold water just before drying. If vegetables are covered with soil, wash them under clean running water to prevent the dirt from resettling on the food. Do not allow vegetables to soak in water. Most vegetables should be peeled and trimmed then cut, sliced, or shredded into uniform pieces. Although peeling some vegetables such as young zucchini and well-washed carrots is optional, unpeeled vegetables tend to be tougher when dried. Remove fibrous or woody portions and damaged areas.

You can prepare pieces with a food slicer or food processor. Although you can dry and store many foods without pretreatment, pretreatment generally improves quality, particularly for vegetables. Five major reasons for treating foods before drying are to:

— Preserve colour and flavour
— Minimise nutrient loss
— Stop decomposition (enzyme action)
— Ensure more even drying
— Extend storage life

DRYING FRUITS AND VEGETABLES

Decomposition from enzyme action during storage is less a problem with fruits than it is with vegetables. Fruits have higher levels of sugar and acid, which counteract enzyme action. Although pretreating fruit is not necessary, you can use an ascorbic acid/citric acid dip, a salt solution dip, syrup blanching, a honey dip, or a sulphiting procedure. Certain fruits, such as apricots, pears, peaches, and some varieties of apples, tend to discolour with drying. Pretreating those fruits can decrease browning during processing and storage and lower losses of flavour and of vitamins A and C.

If you use a pretreatment method that requires soaking fruits in a water solution, you will need to increase drying time because the fruit will absorb some water. Do not allow foods to soak more than 1 hour. Ascorbic Acid/Citric Acid Dips. Ascorbic acid/citric acid dips are often used as a pretreatment for fruits. They prevent fruits such as apples, pears, peaches, and apricots from turning brown when cut and exposed to air. An ascorbic acid dip also increases the vitamin C content of the dried fruit. (Ascorbic acid is another name for vitamin C.). Use U.S.P. ascorbic acid or food-grade ascorbic acid, which are seasonally available among canning supplies in supermarkets.

Vitamin C tablets can also be used. To prepare an ascorbic acid solution, combine ½ teaspoon of ascorbic acid crystals, or three crushed, 500-milligram tablets of vitamin

C, with 1 quart water. Stir until the ascorbic acid dissolves. Place the cut fruit in the ascorbic acid solution. Stir the fruit to ensure even coating. Leave the fruit in the ascorbic acid solution for about 5 minutes. Approximately 1 quart of solution will treat 8 cups of fruit. Pineapple juice or juice from citrus fruits such as oranges, lemons, or grapefruit can also be used as a pretreatment.

These juices contain a mixture of citric and ascorbic acids. However, citric acid is a weaker acid than ascorbic acid and is less effective as a pretreatment. You can also use a commercial pretreatment such as the anti-darkening powders often sold with food preservation supplies. Follow the label directions.

Salt Solution Dip: Prepare a solution of 2 to 4 tablespoons of salt per gallon of water. Soak fruit for 2 to 5 minutes, and then drain it well.

Syrup Blanching: Prepare fruit for drying. Prepare a sugar syrup made with 1 part sugar and 2 parts water. If desired, use less sugar. Bring the syrup solution to a boil. Add the fruit, simmer for 5 minutes, then drain the fruit. Place the fruit on drying trays and dry. This fruit product is like a candied fruit.

Honey Dip: A honey treatment for fruit can effectively minimise browning and softening in light-coloured fruit. Prepare a honey-water dip using 1 part honey to 4 parts water. Dip the fruit in the honey solution immediately after slicing, let it soak for about 5 minutes, and drain well. The dried fruit will have a slight honey taste.

Sulphiting: Sulphur dioxide treatments, either sulphiting or sulphuring, are very effective for retarding oxidation and browning in fruit. Fruit flavour and storage life may also improve. Almost all commercially produced light-coloured fruits, such as dried apples, pears, and apricots, are treated with sulphur compounds. However some people have severe allergic responses to sulphur compounds. They should not eat or work with dried fruit pretreated with sulphur or sulphite compounds.

Sulphuring, a complicated and potentially dangerous procedure, is no longer recommended: Sulphiting involves preparing a solution of water and a sulphiting agent and then soaking the cut fruit in the solution. In the United States six sulphur compounds (sulphur dioxide, sodium sulphite, sodium bisulphite, potassium bisulphite, sodium metabisulphite, and potassium metabisulphite) have been listed by the U.S. Food and Drug Administration (FDA) as "Generally Recognised as Safe" (GRAS).

The most popular sulphiting agents for home drying are sodium bisulphite, sodium sulphite, and sodium metabisulphite. They should be either U.S.P. (food grade) or reagent grade (pure). They are available at most wine-making supply centres and some larger supermarkets.

The sulphiting process has two steps:

— Prepare the sulphiting solution in a large glass container just before use. Place the cut fruit in the solution. Do not leave the fruit in the sulphiting solution too long or the fruit will be mushy. Use about 10 minutes for sliced fruit and 30 minutes for halved fruit. Do not exceed the recommended quantities of sulphites or soak times.

— After sulphiting, remove the fruit and drain it well. Some people recommend a quick rinse in cold water before drying. Place sulphited fruit on drying trays and dry. Drying times for sulphited fruits are longer because the fruit absorbs some water during soaking.

Blanching is the pretreatment method of choice for vegetables. Almost all vegetables should be blanched before drying to destroy the enzymes that make vegetables deteriorate. Blanching keeps vegetables from browning, becoming bitter, or developing off flavours. Blanching also cleans and softens vegetables and makes them easier to rehydrate later. Although you can use either boiling water or steam for blanching, vegetables lose more nutrients during boiling.

Steam Blanching: Use a steamer or make a steamer out of a kettle with a tight-fitting lid. Place a colander, wire basket, or sieve inside the kettle. Make sure the food will be above the water level. Add 2 inches of water to the kettle and heat it to boiling. Place the container with the loosely packed food in the steamer, cover the kettle tightly, and continue boiling.

Water Blanching: Fill a kettle with enough water to cover the food. Bring the water to a rolling boil and gradually stir in the food. Cover the kettle tightly and boil. You can reuse the water when blanching more of the same food, adding more water as necessary. If the water appears dirty, replace it with clean water.

Determining Blanching Times: Blanching times vary with altitude, the type and texture of the vegetable, the amount of vegetable, and the thickness of the pieces. Generally, vegetables should feel and taste firm yet tender. They should not be fully cooked, but they should be heated all the way through. Test the food by cutting through a piece. If sufficiently blanched, it will appear cooked nearly to the centre. Underblanching may cause deterioration in storage, poor rehydration, or bad colour. Overblanching makes vegetables lose colour, flavour, and nutrients and gives them poor texture after rehydration.

After Blanching: Drain vegetables by pouring them directly on the drying trays. If you plan to reuse the water, place a large pan under the trays. Wipe the bottom of the drying tray with a clean towel to remove excess water. Draining the vegetables on one tray and then transferring them to the drying tray results in unnecessary handling. Immediately transfer the blanched vegetables into the dehydrator so drying can begin while the vegetables are still warm.

Different foods can be dried at the same time, but try to choose foods that will dry in about the same amount of time. Onions, peppers, and other strong foods tend to flavour other foods, so dry them separately. Moisture must be removed from the food as quickly as possible at a temperature that does not seriously impair the flavour, texture, or colour of the food. If the temperature is too low at the beginning, the food may spoil before it dries. If the temperature is too high, the surface may harden so that the interior dries much more slowly.

Start the dryer at 140° to 150°F, with the exceptions noted in the drying guidelines. After 2 to 3 three hours, lower the dryer temperature to 130°F to 140°F. Adequate air flow can reduce drying times. Monitor the drying process. If necessary, rotate the trays to ensure even drying. You may need to stir grated, shredded, or finely cut foods.

Drying Time

Many factors affect drying time, including type of food, size and moisture content of the food pieces, pretreatment method, dryer type, dryer temperature, relative humidity of the air, and amount of air movement in the dryer and in the surroundings. With so many factors at work, it's impossible to give precise drying times. Generally, you can figure on drying times of 6 to 36 hours for fruit and 3 to 16 hours for vegetables, which take less time due to their lower sugar contents. In the end, you need to decide when food is dry. Vegetables are sufficiently dry when they are brittle or leathery. Leathery vegetables will be pliable and spring back if folded.

Brittle vegetables such as corn and peas will shatter when hit with a hammer. Fruits are sufficiently dry when they are pliable and leatherlike and have no pockets of moisture. Herbs are sufficiently dry when brittle. Their leaves will shatter when rubbed together. When you think the food is sufficiently dry, remove a piece and allow it to cool completely. Then check for dryness. When you are in doubt about the dryness of a food, continue to dry it. Foods dry more quickly toward the end of the drying period, so check them frequently, and avoid leaving them in the dryer after they are done. Leaving them in will reduce their quality.

REFERENCES

Ashby, B.H., R.T. Hinsch, L.A. Risse, W.G. Kindya, W.L. Craig; and M.T. Turezyn, *Protecting Perishable Foods During Transport by Truck*, HBK-669. USDA, AMS, September, 1987.

Fuell, L. D., D. C. Miller, and M. Chesley, *Dictionary of International Agricultural Trade*, USDA, FAS, June 1988.

Friedheim, Amy, *Ocean Liner Freight Rates and Services*, USDA, AMS, May 1991.

Lipton, W.J. and J.M. Harvey, *Compatibility of Fruits and Vegetables During Transport in Mixed Loads*, MRR-1070. USDA, Fresno, CA, 1977.

McGregor, Brian M., *Tropical Products Transport Handbook*, HBK-668. USDA, AMS, September 1989.

11

Quality Management of Agricultural Crops

Both quantity and quality of crops produced are affected by difficulties in obtaining inputs. The correct planting material is particularly important. Often consumers have strong preferences for particular varieties of vegetable. In much of the Middle and Near East, for example, there is a preference for plum-shaped tomatoes. Red-coloured fruits, such as red sweet apples, are sometimes much preferred to green and golden varieties. Growers'marketing margins can be improved by ensuring the supply of the correct planting material.

The extension officer's role is to advise the nurserymen and seed suppliers on which varieties they should supply, and the farmers on which varieties to plant. Pest and disease damage will seriously reduce a crop's price and its potential shelf life. Sometimes these problems can be solved by the correct crop protection practices. An example of this is the introduction of spray programmes to control scab disease on apples in Kashmir, India.

A crucial step in the successful introduction of this programme was ensuring that agricultural chemical shops had the recommended materials available and could advise growers on their use. The extension officer should be alert for opportunities where farmers themselves can become input suppliers. Farmers can also provide contract services, e.g. for soil cultivation or crop spraying. A critical production constraint is often shortage of funds. Broadly speaking the potential sources of funds can be divided into two-formal and informal. Formal sources are mainly banks.

In general, the interest terms are reasonable, but requirements for security and slow bureaucracy often limit loan effectiveness. The extension officer can help with the supply of production credit by providing the bank with cost-of-production data and likely returns. This will enable the bank to see how much money needs to be advanced. Loan agencies have different rules as to what proportion of the production costs can be advanced. Some will only cover the cost of inputs, while others will include some or all of the labour costs. It is an advantage if the marketing cost of transport and packaging can also be covered.

Formal sources of credit are also extremely valuable to cover long-term investments, such as planting and establishing fruit orchards and investment in production equipment. Some banks offer marketing loans where money is provided to cover the harvesting, transport, packaging and even storage of crops. The amount advanced can be calculated on the basis of the harvesting and marketing costs, as set out above. Sometimes they cover a portion of the estimated wholesale value of production.

The value of these loans is that they can free the grower from the marketing restrictions imposed when borrowing from a middleman. These normally commit the grower to selling all his produce via the one middleman. Where growers are under contract to supply produce to an agribusiness, e.g. food processor or exporter, production loans based on the hypothetical value of the crop can be introduced. For example, if a farmer is growing one acre of tomatoes for a food processor and he would normally harvest five tonnes per acre with a contract price of $0. 25 per kg, his income is expected to be $1 250.

Under a loan hypothecation scheme the bank or processor advances a proportion of the expected income as production credit, say 25 percent or in this case $312.5, without seeking any additional security. Loan recovery can be made from the agribusiness from the money it is returning to the grower. Although informal credit sources include family members, friends and input suppliers, the most important source of credit is often the middleman or commission agent.

The role of a wholesaler as a banker is much misunderstood. Common criticisms are that very high interest rates are charged and that growers who have borrowed money are forced into selling their produce at low prices. In some cases this is true but it is not as common as is assumed. Generally the produce is auctioned in an open market so the grower should receive a fair price.

The loans are often made on the basis of close kinship links and are free of time–consuming bureaucracy. Loans are recovered simply by deducting the money advanced from the sales. The main disadvantage of this system is that the grower has no marketing flexibility. In the case of poor prices he cannot switch to another commission agent or market. In the absence of an effective and suitable formal credit agency, marketing middlemen can be a useful credit source, particularly if finance is a major constraint to starting production. Extension officers should be able to advise farmers on planning their crops. Although important criteria such as labour availability and crop rotations will have to be taken into account the key approach will need to be market oriented production. This means growing crops for which there is likely to be a demand and which will probably be profitable.

The extension officer should calculate potential or net returns of the major alternative horticultural crops in the region. This will establish which crops are likely to be the most profitable. The market research that he or she will have undertaken should have shown whether any of the local crops have a comparative advantage. The produce may have advantages in terms of price, quality or seasonality over competing crops from other areas. Research should also have shown what varieties are favoured and the best time to supply the market. The extension officer will need to translate the market requirements into practical recommendations for farmers, covering areas such as:

— best varieties;
— sowing dates, e.g. whether to extend the period of supply or aim for a particularly high priced period; whether to avoid times of oversupply;
— other techniques to extend production into high priced periods, such as late or early varieties, transplanting techniques, polythene tunnels, irrigation;
— techniques to improve quality, such as optimum fertilisation, crop protection, pruning, irrigation, weather protection.

New crops or the introduction of new technologies or production techniques should always be undertaken, initially on a small-scale trial basis. In practice crops to be planted are chosen for a combination of reasons. The most important is that there is likely to be a good market demand and the crop can be grown profitably. Generally it is advisable to grow a range of crops as this lessens the impact of an eventual crop or market price failure. If possible some crops could be grown under contract. Very often individual growers will have preferences for crops which they feel comfortable growing and/or which grow well on their land.

As agriculture develops farms typically become more specialised, concentrating on fewer crops. Growers become more skilled, although experience has shown that growers can rarely be experts in more than three or four crops. As explained earlier, the most potentially profitable crops are often the most risky. It is useful to have a cropping system in which risky crops are balanced against more reliable income-earning crops. Small farms generally have more labour available per acre. They can take advantage of this by concentrating on growing labour-intensive crops. These are crops which cannot be harvested mechanically and may also require transplanting, pruning, hoeing and multiple-hand harvesting.

Investment Planning

Business and investment advice can also be part of an extension officer's work. Often farmers are tempted to make investments which are expensive and do not improve the viability of their farm significantly. The priorities for investment decisions should be:

— investments that can assure a farmer of an income;
— investments that can increase incomes by improving prices;
— investments that can improve incomes by increasing yields;
— investments which reduce production costs;
— investments which can create additional income sources.

Investments which can improve yield stability are:

— those which protect against pests and diseases, such as sprays and a sprayer,
— those in irrigation and technology which reduce the adverse effects of the weather, e.g. through protection of nursery plants.

Investments like these are particularly important in horticulture because in seasons when yields are low as a result of poor weather or pests, prices rise significantly. It is the growers whose yields arc least affected who make the most profit. Investments which improve prices can be:

— those in technology which facilitate off-season production, such as polythene tunnels and improved transplant technology;
— those made directly in improving marketing such as grading facilities, on-farm storage, pick-up trucks to transport produce and even telephones to improve market communication.

MARKETING ISSUES

The challenge to the agricultural extension officer with special responsibility for marketing is firstly to decide how the marketing problems of the area can be solved. Secondly, he or she needs to think through the best way to get advice or plans across to the maximum number of target farmers. Finally, the agreement and the commitment of those that will be involved in any coordinated production programme in the area must be obtained.

The ways in which problems can be solved and opportunities exploited will change from area to area. To help the extension officer take a clear overview of the area, he or she will need to identify what stage in horticultural development the region has reached. Normally the aim will be to try to introduce the next steps in horticultural progress. The best solutions to marketing problems are normally relatively simple and ought not to require any major changes in production or new technologies. Complex plans or highly innovative plans are much more likely to fail. Three examples are given:

1. If the area produces but has not before sold horticultural crops then the extension officer should be looking for ways to establish a local market. This might involve, for instance, coordinating growers to assemble their produce on one particular day of the week at a convenient location. Middlemen would be invited to attend the market. The buyers would compete with one another to buy produce so that fair prices should be

achieved. As produce is assembled in volume, cheaper bulk transport to the major markets is made possible.

2. If produce is only being sold into a local market there may be opportunities to start supplying more distant, major markets. The extension officer's research should have indicated what the produce requirements, prices and costs are likely to be and identified potential trading partners. He may then want to persuade growers to attempt a test marketing programme to new markets, initially of existing produce and then for new types of produce. If this proves successful then commercially sized shipments could be made. Another option would be for a local representative to take on the role of transporting and selling produce.

3. In an area which is already a major force in horticultural production and marketing the extension officer may concentrate on improving the existing system.. When individual growers already have good links with the market it will be difficult to form producer groups or cooperatives–unless there is some major problem or need. Extension advice is likely to cover generalised advice to groups of growers and specific advice for individual farmers. Potential improvements are likely to be identified at critical points along the production/marketing chain. These may involve anything from new crops and improved production practices to post-harvest techniques, better designed packaging, improved transport methods, better access to credit and production inputs and the establishment of improved market information services.

Finally, it is important to remember that there is always the danger of trying to make changes when they are not necessary. All systems are imperfect inevitably farmers will always complain that they receive too little money while consumers complain of too high prices. Proposals for change do need to be carefully thought through and any additional costs or disadvantages balanced against the advantages.

Extension Officer

The two chief functions of an extension officer are:

— to reduce the learning time for an individual farmer to accept a new idea or technique;

— to increase the number of farmers who understand the new ideas.

In some countries certain extension officers are given the responsibility of becoming subject matter specialists in marketing. It is then their task to train other extension officers and provide specialist marketing advice. By working through other extension officers their effectiveness is increased. Generally the marketing extension officer is based in the production area. He or she must make regular visits to the markets in order to maintain contacts and keep in touch with changes in price and demand. Sometimes, however, he is based away from the production areas.

In this case he will have to return regularly to the production areas to maintain contact with growers' problems and to provide relevant marketing advice. Experience has shown that one of the most effective ways of working is to work with groups of farmers. An extension technique, very much under-exploited, is helping farmers indirectly by providing guidance and advice to private-sector companies. For example, companies who either supply inputs or, more importantly, assist in the marketing and distribution of produce can often use timely and reasoned advice to put into motion a process which brings benefits to large numbers of growers.

Leading local farmers strongly influence the decisions of other farmers, e.g. to coordinate their production and marketing. They will generally need to be persuaded of the potential advantages to themselves of improved marketing practices. The extension officer should beware that unless the wealthy farmer is genuinely concerned for the community's welfare he may try to suppress the planned changes. He may do this because he does not want other farmers to become better off. Alternatively, he may see an advantage in carrying out the scheme on his own.

The scope of the work clearly extends far beyond advising farmers. This is because marketing is only as successful as the weakest link in the production/ marketing chain. The first involves giving advice to an individual farmer. This should be a low priority as it is an inefficient use of time and larger-scale farmers obtain the most benefits. The second involves providing marketing advice to farmer groups, particularly through the methods of mass extension. This has the advantage of reaching a large number of growers and allows coordination of farmers' activities and cooperation in marketing.

The third technique involves providing advice or information to critical individuals, organisations or private-sector companies in the marketing chain whose actions can have a beneficial effect on marketing. Finally, and perhaps most ambitiously, an extension officer may decide it is necessary to attempt a project approach to developing horticultural marketing. Marketing is normally achieved by a series of interlinking stages and coordination between the stages is essential. A project approach is one which involves coordinating the activities of a number of different intermediaries in a marketing chain. It may involve a group of farmers assembling their produce at one point so that it can be transported in bulk to the market.

More complex schemes could involve ensuring a supply of inputs, providing growers with production advice and negotiating contract terms with a buyer, be he a food processor or exporter. Extension officers can have a credibility problem. One way of overcoming this is to achieve a good reputation by successfully resolving some smaller problems. The horticultural marketing advisor achieved credibility by negotiating significantly cheaper air freight rates. A second way involves securing influential support for the scheme, particularly from farmer leaders or marketing companies. Sometimes them is a reluctance for people to implement someone else's ideas enthusiastically.

A clever advisor sometimes overcomes this problem by not revealing his project plans to the individual he considers to be the most important in the scheme. Instead he provides him with the information on the problem and then in discussion leads the individual to come up with the same (or a similar) solution. The individual then thinks it is his own idea and has the enthusiasm and commitment to ensure the plan's implementation.

Some of the ideal marketing techniques are listed below:

— *Farmer explains farmers*: A successful farmer explains to a group of farmers his production and marketing practices. The meeting is most effective on the farmer's own farm.

— *Practical demonstrations techniques*: Practical demonstrations of techniques such as harvesting, cleaning, grading and packing, preferably taking place on a farm. Prepared samples which demonstrate the differences overtime of different handling practices can be effective, as are samples of competing produce and photographs.

— *Talks and seminars:* Possible topics include market possibilities successful case studies, post-harvest techniques, prigs assessment, market-oriented production techniques. Buyers and middlemen should be involved to talk.

— *Problem-solving techniques:* The farmer group is encouraged to identity its own major problems. The problem solving can be tackled systematically, by calling In specialists individually to advise the group or by forming a panel to answer farmers' questions. Alternatively, the group might be encouraged to decide their own solutions which they then implement themselves collectively.

— *Study tours*: Farmers are taken on a study tour to make their own contacts and to see the market for themselves, visit processing centres and observe how their produce withstands transportation. Farmers visit farmers in another area to exchange experiences and see new techniques. This experience alone can transform a grower's views on production and marketing.

— *Written information*: Fact sheets are prepared and distributed. These can identity potential trading partners or provide technical information on production and post-harvest techniques.

— *Market news services*: Establishing a market news service which provides regular, reliable, relevant and timely information. This may be in the form of a news sheet or a radio bulletin.

The extension officer's role is then to support the individual, to coordinate the activities of the different parties involved and to chase up the progress of the project. A planned project approach to horticultural marketing development will increase the chances of genuine improvements being made. It is important for the extension officer to have a clear mental image of the desired outcome and successfully communicate that objective.

The project must be understood by all parties, if they are going to be able to work effectively together. Business management experience has shown that targets, such as tonnages to be shipped or selling prices, are important too, as they provide challenges to the parties involved and can be used to monitor the progress of the project. This technique is called management by objectives.

However, no matter how good the preliminary work has been, when a plan is put into action the unexpected will happen. Allowances for the unexpected should be made. It is advisable to start the project with a pilot stage so that mistakes can be made on a small scale from. Furthermore the project must be flexible so that changes can be made in the light of these lessons. Inevitably the project's critics will try to emphasise any problems; most successful projects will have had to face problems, particularly in the early phases, and their success is often a measure of their ability to learn from and overcome difficulties.

ORGANIC AGRICULTURE

Organic agriculture is best known as a farming method where no synthetic fertilisers and pesticides are used. It is a holistic production management system which promotes and enhances agro-ecosystem health, including biodiversity, biological cycles and soil biological activity. It emphasises the use of management practices in preference to the use of off-farm inputs, taking into account that regional conditions require locally adapted systems. This is accomplished by using, where possible, agronomic, biological and mechanical methods, as opposed to using synthetic materials, to fulfil any specific function within the system.

Several national governments and a multitude of private certification and farmer organisations have defined organic agriculture. One of the essential elements distinguishing organic farming from other forms of sustainable agriculture is the existence of production standards and certification procedures. There are no universal standards for production and handling of organic fruit and vegetables. Initially, organic standards were developed by private associations, entitling members to use the respective associations' organic brands and labels when marketing their products.

The International Federation of Organic Agriculture Movements (IFOAM), a non-governmental organisation promoting organic agriculture internationally, has established guidelines that have been widely adopted for organic production and processing. As organic agriculture has become more widespread, many developed countries have defined their own organic standards. Since the early 1990s, EC countries have endorsed a common organic standard which is spelled out in Regulation EEC 2092/91. More recently, Canada, the United States and Japan have adopted organic standards and regulations.

The Committee on Food Labelling of the FAO/WHO Codex Alimentarius Commission adopted "Guidelines for the Production, Processing, Labelling and Marketing of Organically Produced Foods" in 1999. Organic standards are usually similar as they derive from IFOAM's guidelines for organic production. In general, the use of methods that contribute to maintaining or enhancing soil fertility is mandatory. Another common feature is that generally natural inputs are approved and synthetic inputs are prohibited. Nevertheless, there are exceptions in both cases. Certain natural inputs which various certification programmes determine to be harmful to human health or the environment are prohibited.

In addition, certain synthetic inputs are allowed. For example, EEC Regulation 2092/91 allows, when required, the use of specific fertilisers and plant preservatives. All certification programmes maintain lists of specific approved synthetic inputs and prohibited natural inputs. Many certification programmes require additional environmental protection measures beyond the above prerequisites. For example, specific measures are generally applied in the areas of soil and water conservation, pollution control or the use of biological control agents.

While many farmers in the developing world do not use synthetic inputs, this alone is not sufficient to classify their products as organic. Farmers who produce organic crops for export and at the same time grow basic foodstuffs on other areas of the farm using conventional methods with fertilisers and pesticides, inadmissible to the organic system, are at high risk of violating the standards, unless effective measures are taken to prevent prohibited substances from passing to the organic plots.

Producers and exporters of organic fruit and vegetables seeking to sell their products in developed countries which have adopted organic standards and regulations will have to meet the rules established by the importing country concerned. In the United States, Japan and the EC countries, the largest markets for organic products, the standards applying to domestic organic production also apply to imported organic products. There are detailed regulations governing the production, importing, marketing and labelling of organic products. Producers and exporters wishing to export fruit and vegetables with the organic label will have to obtain organic certification. Organic certification is a procedure for verifying that the production process conforms to certain standards. In other words, certification is primarily an acknowledgement that these products have been produced according to organic production standards.

The basic role of a certification body is to confirm that these adhere to a specific organic standard. Once certified, organic products are marketed, usually carrying a certification label, indicating that the products are certified as organic. The certification label attests to conformity with certain standards and is itself not a trade mark. However, in most countries the certification label is also registered as a trade mark. Producers wishing to enter a specific market for organic products may find it useful to be certified by a certification body with a certification label that is well known in that market.

The cost of certification can be high, although it varies in relation to farm size, volume of production and which certification body is chosen. Relatively few developing countries have certification bodies within their borders, and even when sufficient resources are available to pay for certification, farmers often lack the information to find credible inspectors. To date, government regulations essentially deal with references to the organic production method on labels. They do not define standards for the use of private trade marks or certification marks. This implies that government regulations may be supplemented by private quality seals based on private organic standards.

ORGANIC PRODUCTION SYSTEM

Organic agriculture is a production management system that aims to promote and enhance ecosystem health, including biological cycles and soil biological activity. It is based on minimising the use of external inputs, and represents a deliberate attempt to make the best use of local natural resources. Methods are used to minimise pollution of air, soil and water, although they cannot ensure that products are completely free of residues, because of general environmental pollution.

Organic agriculture comprises a range of land, crop and animal management procedures. Unlike food labelled as "environmentally friendly", "natural" or "free-range", organic agriculture is circumscribed by a set of rules and limits, usually enforced by inspection and certification mechanisms. Other terms used, depending on the language, are "biological" or "ecological". "Biodynamic" refers to commodities that are produced according to organic and other additional requirements. Synthetic pesticides, mineral fertilisers, synthetic preservatives, pharmaceuticals, GMOs, sewage sludge and irradiation are prohibited in all organic standards.

Plant nutrient or pesticide inputs derived directly from natural sources are generally allowed, as is a minimum of pretreatment before use. Industrially produced pesticides, for example, may not be applied in organic agriculture, but an extract of neem leaves, which have biocidal properties, is currently allowed. Most industrial countries, but few developing countries, have national organic standards, regulations and inspection and certification systems that govern the production and sale of foods labelled as "organic". At the international level, the general principles and requirements applying to organic agriculture are defined in the Codex guidelines adopted in 1999.

The growing interest in organic crop, livestock and fish products is mainly driven by health and food quality concerns. However, organic agriculture is not a product claim that organic food is healthier or safer, but rather a process claim intending to make food production and processing methods respectful of the environment. Organic agriculture is not limited to certified organic farms and products only. It also includes non-certified ones, as long as they fully meet the requirements of organic agriculture. This is the case for many non-certified organic agricultural systems in both developing and industrial countries where produce is consumed locally or sold directly on the farm or without labels.

The extent of these systems is difficult to estimate since they operate outside the certification and market systems. Organic practices that encourage soil biological activity and nutrient cycling include: manipulation of crop rotations and strip cropping; green manuring and organic fertilisation; minimum tillage or zero tillage; and avoidance of pesticide and herbicide use. Research indicates that organic agriculture significantly increases the density of beneficial invertebrates, earthworms, root symbionts and other micro-organisms. Properly managed organic agriculture reduces or eliminates water pollution and helps conserve water and soil on the farm. Some countries compel or subsidise farmers to use organic techniques as a solution to nitrate contamination in groundwater.

Organic Management of Land

Growth rates of land under organic management are impressive in western Europe, Latin America and the United States. Between 1995 and 2000, the total area of organic land tripled in western Europe and the United States. In the United States, land under certified organic agriculture has been growing by 20 percent p.a. since 1989, while in western Europe average annual growth rates have been around 26 percent since 1985.

In 1999 alone, the United Kingdom experienced a 125 percent growth of its organic land area. However, these dramatic increases must be viewed against the small starting base levels. In some cases they may reflect a reclassification of land rather than an actual switch in farming systems. Policy measures were instrumental in persuading small farmers to convert to organic farming by providing financial compensation for any losses incurred during conversion. The guidelines established by the organic agriculture community in the 1970s were formalised by national and supranational legislation and control systems.

The role of organic agriculture in reaching environmental policy objectives, including sustainable use of land set-aside, led to the adoption of agri-environmental measures that encourage organic agriculture. The estimates given below are derived from a compilation of available information by the Foundation Ecology and Agriculture (SOL) in Germany.

In the absence of a statistical database on organic agriculture in FAO, this is the most comprehensive source available. SOL reports a global area of land under certified organic agriculture of 15.8 million ha of which 48.5 percent are located in Oceania (Australia 7.7 million ha); 23.5 percent in Europe (with Italy having the highest area, nearly 1 million ha); 20 percent in Latin America (with Argentina having 3 million ha); 7.4 percent in North America (United States nearly 1 million ha); 0.3 percent in Asia; and 0.1 percent in Africa.

With 3 million ha, Argentina accounts for more than 90 percent of the certified organic land in Latin American countries and has the second largest area of organically managed land in the world after Australia. In both countries, however, most of this is grassland. Because of the large size of organic ranches in the pampas, the average size of organic farms is 3000 ha in Argentina. Some 85 percent of Argentina's organic production is exported.

In 2000, agricultural land under certified organic management averaged 2.4 percent of total agricultural land in western Europe, 1.7 percent in Australia, 0.25 percent in Canada and 0.22 percent in the United States. In most developing countries, agricultural land reported under certified organic production is minimal (less than 0.5 percent of agricultural lands). However, some traditional farms in developing countries have adopted modern organic management to improve their productivity, especially in areas where pesticides and fertilisers are inaccessible.

The extent of such non-market organic agriculture is difficult to quantify but some attempts have been made. The Ghanaian Organic Agriculture Network, for example, estimates that there are around 250000 families in South and East Africa farming around 60 million ha on an organic basis. Over one-third of West African agricultural produce is produced organically. A number of industrial countries have action plans for the development of organic agriculture. Targets are set for the sector's growth and resources are allocated to compensate farmers during, and sometimes after, the conversion period, and to support research and extension in organic agriculture.

For example, the United Kingdom increased the budget of the Organic Farming Scheme to support conversion to organic agriculture by 50 percent for 2001-02 (£20 million per year) and allocations in the United States for the organic sector include US$5 million for research in 2001. India and Thailand have established their own organic standards to facilitate exports and to satisfy domestic demand. China, Malaysia and the Philippines are at present working towards establishing national standards.

Profitability

Typically, farmers experience some loss in yields after discarding synthetic inputs and converting their operations from conventional, intensive systems to organic production. Before restoration of full biological activity, pest suppression and fertility problems are common. The degree of yield loss varies and depends on inherent biological attributes of the farm, farmer expertise, the extent to which synthetic inputs were used under previous management and the state of natural resources. It may take years to restore the ecosystem to the point where organic production is economically viable.

Transition to organic management is difficult for farmers to survive without financial compensation, especially in high intensive input agriculture and in degraded environments. After the conversion period, organic agriculture achieves lower yields than high external input systems. Depending on the previous management level of specialisation, yields can be 10 to 30 percent lower in organic systems, with a few exceptions where yields are comparable in both systems.

In the medium term, and depending on new knowledge, yields improve and the systems' stability increases. In the longer term, performance of organic agriculture increases in parallel with improvements in ecosystem functions and management skills. Yields do not usually fall, however, when conversion to organic agriculture starts from low-input systems (often traditional) that do not apply soil-building practices. A study from Kenya indicates that, contrary to general belief, organic agriculture in the tropics is not constrained by insufficient organic material (to compensate for the non-use or reduced use of external inputs), but instead shows a good performance.

Average fertiliser consumption will rise in developing countries. The average figure masks, however, that for many (especially small) farmers the purchase of manufactured fertilisers and pesticides is and will continue to be constrained by their high costs relative to output prices or simply by unavailability. Organic agriculture emphasises understanding and management of naturally occurring production inputs as an alternative to enhance yields. It will not be possible for organic agriculture to attain the high yields achieved with the use of synthetic inputs in high-potential areas. But organic management offers good prospects for raising yields and the sustainability of farming in resource-poor and marginal areas, and can raise the productivity of traditional systems while relying on local resources.

For example, India is collecting nationwide information regarding the experiments being carried out in organic agriculture, with a view to reintroducing it as part of its traditional "rishi agriculture". In Latin America, hundreds of thousands of indigenous farmers along the Andes have turned to the organic movement to reinstate sophisticated agricultural practices developed by the Incas. Individual small family vegetable plots and groups/associations managing organic produce for domestic urban markets and small informal fairs are widespread.

Cuba adopted organic agriculture as part of its official agricultural policy, with investments in research and extension, to compensate for shortages in external inputs. In 1999 (non-certified) organic urban agriculture (in self-provision gardens, raised container beds and intensive gardens) produced 65 percent of the country's rice, 46 percent of fresh vegetables, 38 percent of non-citrus fruit, 13 percent of roots, tubers and plantains and 6 percent of eggs. In organic systems, external inputs such as fertilisers, herbicides and machinery are replaced by labour, most often increasing women's work. Labour can either be a major constraint to organic conversion, or an employment provider to rural communities. Often the introduction of organic agriculture shifts gender distribution of labour as men prefer to be involved with mechanised agriculture.

Women rarely own land and are dependent on access to common property. Since access to credit frequently requires land as collateral, women (and landless people) are largely excluded from the formal credit market. As a result, women seek methods that require little external inputs. Organic agriculture facilitates women's participation as it does not rely on financial inputs and access to credit. The economic performance of organic agriculture in industrial countries (mainly in Europe) is determined by financial support from governments, premium prices for produce and high labour costs.

An extensive analysis of European farm economics in terms of labour use, yields, prices, costs and support payments, concludes that profits on organic farms are, on average, comparable to those on conventional farms. However, only a few studies have assessed the long-term profitability of organic agriculture. Profitability of organic systems relates to whole-farm production (total production of a variety of species and not single crop yields) over the entire rotation period. This includes both marketed products and non-food production.

Incomes achieved over a given season may appear high because of price premiums when excluding the low profits over rotational seasons. Supply constraints faced by organic farmers, which are expected to increase if the sector expands, include the provision of adequate organic inputs such as organic seeds, natural pest enemies and mineral rocks. An increased organic food supply above a certain level would lead to a decline in premium prices.

Organic Food Market

On the demand side, promotion and marketing strategies of retailers and supermarkets, in particular of major food-retailing chains, have created new market opportunities for organic agriculture in industrial countries. Food-retailing chains, which also stock and promote organic foods as a tool to improve their public image, account for a major share of the retail markets for fresh as well as processed organic foods. Concerns about growth-stimulating substances, GM food, dioxin-contaminated food and livestock epidemics have given further impetus to organic food demand as consumers increasingly question the safety of conventional foods.

The most recent outbreak of foot-and-mouth disease has added to concerns over the soundness of industrial agriculture. Several governments have responded with declarations of targets for the expansion of organic production. Many consumers perceive organic products as safer and of higher quality than conventional ones. These perceptions, rather than "science", drive the market.

The market opportunities arising from these concerns have also opened possible niche markets for developing country exporters. Major industrial countries' markets offer good prospects for suppliers of organic products that are not produced domestically as well as off-season products and processed foods. Liberalisation and privatisation policies in developing countries open the way for a greater role for organic entrepreneurs and producers' organisations. Markets for value-added products such as organic commodities can help counterbalance falling commodity prices and withdrawal of government support for agricultural inputs and other services.

Price premiums range from 10 to 50 percent over prices for non-organic products. There is also government support for organic exports. Examples include the organic coffee programme of the Coffee Development Authority in Uganda; the promotion of organic exports by India's Ministry of Commerce; and support by the Argentinean government for the export of over 80 percent of the country's organic produce. The size of domestic organic production is not necessarily related to the importance of organic markets. Australia, which has the world's largest area of organic land, most of which is grassland, has a market of US$170 million of organic food retails.

Japan, on the other hand, which has only 5 000 ha of organic lands, is the second largest world organic market. The largest markets of organic foods are in western European countries, Japan and the United States. The UNCTAD/WTO International Trade Centre (ITC) estimated retail sales of organic foods in the largest markets at US$20 billion in 2000, of which US$8 billion in Europe and the United States each, and US$2.5 billion in Japan. In spite of dramatic growth rates, sales of organic agricultural products in industrial countries in 2000 represented less than 2 percent of total food sales at the retail level. However, in particular countries and for particular products, the market share of organic agricultural products can be appreciably larger.

Organic food sales in Germany are 3 to 4 percent of total sales, while individual commodities such as organic milk products have over a 10 percent market share, with the figure for organic ingredients in baby foods in the range of 80 to 90 percent. Organic coffee, which accounts for 0.2

percent of world coffee consumption, accounts for 5 percent of the United States coffee market. Some 100 developing countries produce organic commodities in commercial quantities, most of which are exported to industrial countries. Where they exist, developing countries' organic markets are still very limited and food is sold mainly in specialised stores in large cities. ITC estimates an annual sales growth of organic food between 10 and 40 percent over the medium term, depending on the market. Thus, organic food retail sales could grow from an average of 2 percent of total sales in 2000 to a share of 10 percent in major markets in a few years.

REFERENCES

Darke, Rick & Mark Griffiths, eds. *The New Royal Horticultural Society Dictionary. Manual Of Grasses*, London: Macmillan, 1994.

Paul, Tessa, *New Flowers. Growing New Garden Varieties*. N.Y.: Abrams, 1990

Rice, Laura Williams & Robert P. Rice, Jr. *Practical Horticulture*. Englewood Cliffs, NJ: Prentice Hall Career & Technology, 1993

Synge, Patrick M., *A Diversity In Plants*. London: Geoffrey Bles, 1953.

Wyman, Donald, *Wyman's Gardening Encyclopedia*. NY: Macmillan, 1972

Zomlefer, Wendy B., *Guide to the Flowering Plant Families*. Chapel Hill: University of North Carolina Pr., 1994.

Bibliography

Ashby, B.H., R.T. Hinsch, L.A. Risse, W.G. Kindya, W.L. Craig; and M.T. Turezyn, *Protecting Perishable Foods During Transport by Truck*, HBK-669. USDA, AMS, September, 1987.

Chadha, K.L., Ravindran P.N. and Leela Sahijram, *Biotechnology in Horticultural and Plantation Crops*, New Delhi, Malhotra Publishing, 2000.

Cox, Peter A., *The Larger Rhododendron Species*. Portland: Timber Press, 1990.

Darke, Rick & Mark Griffiths, eds. *The New Royal Horticultural Society Dictionary. Manual Of Grasses*, London: Macmillan, 1994.

Dole, J.M. and H.F. Wilkins, *Floriculture principles and species*, 2 nd Edition. Pearson Education, Inc., Upper Saddle River, N.J. Available in Branson Library, 2005.

Fuell, L. D., D. C. Miller, and M. Chesley, *Dictionary of International Agricultural Trade*, USDA, FAS, June 1988.

Hay, Roy & Patrick M. Synge *The Colour Dictionary Of Garden Plants. With House and Greenhouse Plants.*, London: Bloomsbury Books, 1991.

Jacob John, P., *A Handbook on Post Harvest Management of Fruits and Vegetables*, Delhi, Daya Publishing House, 2008

Lipton, W.J. and J.M. Harvey, *Compatibility of Fruits and Vegetables During Transport in Mixed Loads*, MRR-1070. USDA, Fresno, CA, 1977.

McGregor, Brian M., *Tropical Products Transport Handbook*, HBK-668. USDA, AMS, September 1989.

Nicholas, C. J., *Export Handbook for U.S. Agricultural Products*, HBK-593. USDA, AMS, March 1985.

Paul, Tessa, *New Flowers. Growing New Garden Varieties*. N.Y.: Abrams, 1990

Rai N. and Yadav, D.S., *Advances in Vegetable Production*, New Delhi, Researchco Book Centre, 2005.

Rice, Laura Williams & Robert P. Rice, Jr. *Practical Horticulture*. Englewood Cliffs, NJ: Prentice Hall Career & Technology, 1993

Romesh Kumar Sud and Sudhir Kumar, *Flowers and Vegetables of India*, Jodhpur, Scientific, 2004.

Saroj, P.L. et.al., *Advances in Arid Horticulture: Present Status*, International Book Distributing Co., Lucknow, 2004.

Saroj, P.L. Vashishtha, B.B. Dhandar, D.G., *Advances in Arid Horticulture : Vol. I : Present Status*, Lucknow, International Book Distributing Co., 2004,

Seeley, J.G., *Advancements in Commercial Floriculture*, Southern Florist and Nurseryman, 1982.

Sharma, R.R., *Fruit Production : Problems and Solutions*, Lucknow, International Book Distributing Co., 2006.

U.S. Department of Agriculture, Foreign Agricultural Service, *Food & Agricultural Export Directory 1991*. MPN-1497, September 1991.

Utpal Banerjee, *A Handbook of Practical Horticulture*, Jaipur, Mangal Deep, 2002.

Vishal Nath, Dinesh Kumar and V. Pandey, *Fruits for the Future*, Vol. I. *Well Versed Arid and Semi Arid Fruits*, Delhi, Satish Serial Pub., 2008.

Wyman, Donald, *Wyman's Gardening Encyclopedia*. NY: Macmillan, 1972

Zomlefer, Wendy B., *Guide to the Flowering Plant Families*. Chapel Hill: University of North Carolina Pr., 1994.

Index

G.S
17/7/09

1

SAI